Gephi Cookbook

Over 90 hands-on recipes to master the art of network analysis and visualization with Gephi

Devangana Khokhar

BIRMINGHAM - MUMBAI

Gephi Cookbook

First published: May 2015

Production reference: 1220515

Published by Packt Publishing Ltd.
Livery Place
35 Livery Street
Birmingham B3 2PB, UK.

ISBN 978-1-78398-740-5

www.packtpub.com

Credits

Author

Devangana Khokhar

Reviewers

Palash Goyal

Kyunghoon Kim

Pablo Paredes Navarro

Doug Needham

Steven Paul Sanderson II

Commissioning Editor

Ashwin Nair

Acquisition Editors

Purav Motiwalla

Rebecca Youé

Content Development Editor

Kirti Patil

Technical Editor

Ankita Thakur

Copy Editors

Sonia Michelle Cheema

Stephen Copestake

Project Coordinator

Nidhi Joshi

Proofreaders

Stephen Copestake

Safis Editing

Indexer

Monica Ajmera Mehta

Graphics

Sheetal Aute

Production Coordinator

Arvindkumar Gupta

Cover Work

Arvindkumar Gupta

About the Author

Devangana Khokhar is a consultant at ThoughtWorks Inc., working on a range of exciting projects, primarily in the data science and analytics domain and is currently based out of Bengaluru. She has more than 4 years of experience in data analytics, social networks analysis, machine learning, and information retrieval. She is also the director of Women Who Code's Bangalore chapter, a nonprofit organization focused on bringing more women into the field of technology. She holds a master's degree in theoretical computer science and has specialized in social network analysis from PSG College of Technology, Coimbatore. During her postgraduate study, she was intrigued by social networks and machine learning, and she has been in love with data science and analytics since then. Devangana has also been one of the reviewers for *R Graphs Cookbook Second Edition*, *Jaynal Abedin* and *Hrishi V. Mittal, Packt Publishing*.

She is passionate about spreading the message of educational equality and is an advocate of women's right to education and equal stature in the tech industry. She also takes an interest in cooking and reading books, mostly in the realm of nonfiction. She is a Twitter addict and very often shares resources that she finds interesting or useful in her pursuit of getting better at data science. She tweets at `http://www.twitter.com/DevanganaK`. You can also get in touch with her on LinkedIn at `in.linkedin.com/in/devangana`.

Acknowledgments

Prima facie, I am extremely grateful to my parents for being so understanding and supportive of my passion. If it weren't for their constant words of encouragement and belief in me, my dream of authoring my first book would have never come true. This book is dedicated to them, whose patience, love, and support enabled every keystroke.

I wish to sincerely thank my colleagues and friends who've been with me as my pillars of strength all the time. Their unceasing support kept me motivated throughout this journey.

I would like to extend a big thank you to the reviewers of my book and the various editors from Packt Publishing whose tremendous support and suggestions helped me in getting this book in the best possible shape.

Let me put on record my sense of gratitude to one and all for directly or indirectly lending their support to this venture.

About the Reviewers

Palash Goyal graduated from IIT Guwahati and with a specialization in mathematics and computing. He is passionate about data science and its applications and spends most of his time researching on data science techniques. He has experience in digital marketing, web analytics, and mathematical finance. In his free time, he likes to travel, sketch, and follow blogs related to upcoming technologies and tech-products.

He is presently working with the data science team of MakeMyTrip Pvt. Ltd. and has experience working with start-ups in online marketing on Facebook and Google platforms as well as Oracle as an application developer. You can find his data science, neural networks, and artificial intelligence related tweets at `http://www.twitter.com/palgoyal1` and you can connect with him on LinkedIn at `in.linkedin.com/in/palashgoyal1`.

> I would like to thank my parents, sisters, and friends for their continuous support and help while reviewing this book.

Kyunghoon Kim is the cofounder of the event explorer Core.Today and is a unified MS/PhD student in the Department of Mathematical Sciences at Ulsan National Institute of Science and Technology (UNIST). He is interested in the development of new algorithms that are inspired by nature. Currently, he is focused on researching mathematical programming for data mining, especially its complex network structure. He is also currently working on a long-term project to design the framework to realize the automated mathematician.

Kyunghoon is an avid Python programmer. He talked on NetworkX, a Python library to study networks and graphs, at PyCon, Korea, in 2014.

> I would like to thank the Packt Publishing team for giving me the chance to review this book, and I am pleased that this opportunity came my way. I would also like to thank my professor, Bongsoo Jang, who introduced me to the world of applied mathematics and supported me in all my endeavors. I am also extremely grateful for the support extended by my family and friends, who mean the world to me.

Pablo Paredes Navarro has a bachelor in business administration degree from Adolfo Ibáñez University, Chile, and an MSc in science and technology studies from the University of Edinburgh, UK. He has participated in different research projects, focusing on social network analysis and digital development, and has also talked at length on these subjects at seminars conducted in Chile and the United Kingdom. His current interests are related to the use of SNA for digital networks, e-government, and ICT for development purposes in general.

Doug Needham has been building and managing data-focused enterprises for over 20 years. Distilling data down to a graph and providing various analysis results to be used in customer sites has been his most recent focus. Doug can be reached on Twitter @dougneedham.

Steven Paul Sanderson II is currently in the last year of his graduate program, after which he will obtain his master's degree in public health from the State University of New York at Stony Brook University School of Medicine. He has worked in a hospital setting for about 9 years in various departments. Steven is an active user of StackExchange sites and his aim is to self-learn several topics, most notably Stack Overflow, where he is working to gain an understanding and become a better user of SQL, R, VB, and Python.

He is currently employed as an EpicCare analyst for NYU Langone Medical Center, located in Manhattan, a large forward thinking academic medical center.

Steven has also worked on *Network Graph Analysis and Visualization with Gephi*, *Ken Cherven, Packt Publishing*, and has also coauthored a book with a former professor, Phillip Baldwin, called *The Pleistocene Re-Wilding of Johnny Paycheck*, which can be found as a self-published book at LULU.com: `http://www.lulu.com/shop/phillip-baldwin/the-pleistocene-re-wilding-of-johnny-paycheck/paperback/product-21204148.html`.

I would like to thank Christina for making me complete on March 14, 2014, and being kind enough to become my wife; I love you.

www.PacktPub.com

Support files, eBooks, discount offers, and more

For support files and downloads related to your book, please visit www.PacktPub.com.

Did you know that Packt offers eBook versions of every book published, with PDF and ePub files available? You can upgrade to the eBook version at www.PacktPub.com and as a print book customer, you are entitled to a discount on the eBook copy. Get in touch with us at service@packtpub.com for more details.

At www.PacktPub.com, you can also read a collection of free technical articles, sign up for a range of free newsletters and receive exclusive discounts and offers on Packt books and eBooks.

https://www2.packtpub.com/books/subscription/packtlib

Do you need instant solutions to your IT questions? PacktLib is Packt's online digital book library. Here, you can search, access, and read Packt's entire library of books.

Why Subscribe?

- ▶ Fully searchable across every book published by Packt
- ▶ Copy and paste, print, and bookmark content
- ▶ On demand and accessible via a web browser

Free Access for Packt account holders

If you have an account with Packt at www.PacktPub.com, you can use this to access PacktLib today and view 9 entirely free books. Simply use your login credentials for immediate access.

Table of Contents

Preface

Gephi Cookbook is a guide to learn about interactive network exploration and visualization accompanied by the graph theory concepts that drive them. It helps you to understand about the nuances of network visualization, not only from a conceptual, but also from an implementation perspective. This book is an invaluable resource if you are looking forward to getting a deep-dive into the network analysis domain without having to learn how to code.

What this book covers

Chapter 1, Getting Started with Gephi, will take you through the process of installing Gephi on various platforms. It also gives you an overview of Gephi's GUI and basic understanding of various modes available in it.

Chapter 2, Basic Graph Manipulations, teaches you to perform basic graph manipulations such as adding and deleting nodes, editing node attributes, and applying filters on networks by exploiting the user friendly interface of Gephi.

Chapter 3, Using Graph Layout Algorithms, explores the basic default layout algorithms available in Gephi from both a conceptual as well as an implementation perspective.

Chapter 4, Working with Partition and Ranking Algorithms, will take you through the processes of the ranking and partitioning of graphs based on user-defined metrics and modifying the graph visualization based on various parameters.

Chapter 5, Running Metrics, Filters, and Timelines, will enable you to learn about the statistical properties of graphical networks and how they can exploit these properties with the help of Gephi.

Chapter 6, Working in the Data Laboratory Mode, thoroughly describes the Data Laboratory mode in Gephi, and explores a number of tasks that can be accomplished with the help of this mode.

Chapter 7, Getting Graphs and Networks Ready for Preview, covers the in-built rendering settings of Gephi and the process of exporting the final graph to multiple formats.

Chapter 8, Exploring Dynamic and Multilevel Graphs, focuses on two special kinds of graphs, dynamic graphs and multilevel graphs, and describes their working in detail.

Chapter 9, Getting Real-world Graph Datasets, explores various networks in Gephi. Also, it describes the art of fetching data from a number of different sources.

Chapter 10, Exploring Some Useful Gephi Plugins, describes a number of plugins that are extensively used by researchers and developers while working with Gephi.

What you need for this book

To run the various recipes in this book, Gephi version 0.8 is required. Unless otherwise mentioned, it is best to have the latest version of this software.

Who this book is for

If you want to learn network analysis and visualization along with graph concepts from scratch, then this book is for you. This is ideal for those of you with little or no understanding of Gephi and this domain, but will also be beneficial for those interested in expanding their knowledge and experience.

Sections

In this book, you will find several headings that appear frequently (Getting ready, How to do it, How it works, There's more, and See also).

To give clear instructions on how to complete a recipe, we use these sections as follows:

Getting ready

This section tells you what to expect in the recipe, and describes how to set up any software or any preliminary settings required for the recipe.

How to do it...

This section contains the steps required to follow the recipe.

How it works...

This section usually consists of a detailed explanation of what happened in the previous section.

There's more...

This section consists of additional information about the recipe in order to make the reader more knowledgeable about the recipe.

See also

This section provides helpful links to other useful information for the recipe.

Conventions

In this book, you will find a number of text styles that distinguish between different kinds of information. Here are some examples of these styles and an explanation of their meaning.

Code words in text, database table names, folder names, filenames, file extensions, pathnames, dummy URLs, user input, and Twitter handles are shown as follows: "Open the gephi.conf file and, in the default options, change the -J-Xmx512m value to -J-Xmx1024m."

New terms and **important words** are shown in bold. Words that you see on the screen, for example, in menus or dialog boxes, appear in the text like this: "The upper-left corner on the screen has three tabs namely **Overview**, **Data Laboratory**, and **Preview**, which represent the three modes present in Gephi for network manipulation."

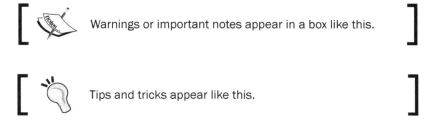

[Warnings or important notes appear in a box like this.]

[Tips and tricks appear like this.]

Reader feedback

Feedback from our readers is always welcome. Let us know what you think about this book—what you liked or disliked. Reader feedback is important for us as it helps us develop titles that you will really get the most out of.

To send us general feedback, simply e-mail `feedback@packtpub.com`, and mention the book's title in the subject of your message.

If there is a topic that you have expertise in and you are interested in either writing or contributing to a book, see our author guide at `www.packtpub.com/authors`.

Customer support

Now that you are the proud owner of a Packt book, we have a number of things to help you to get the most from your purchase.

Downloading the color images of this book

We also provide you with a PDF file that has color images of the screenshots/diagrams used in this book. The color images will help you better understand the changes in the output. You can download this file from `http://www.packtpub.com/sites/default/files/downloads/7405OS_ColorImages.pdf`.

Errata

Although we have taken every care to ensure the accuracy of our content, mistakes do happen. If you find a mistake in one of our books—maybe a mistake in the text or the code—we would be grateful if you could report this to us. By doing so, you can save other readers from frustration and help us improve subsequent versions of this book. If you find any errata, please report them by visiting `http://www.packtpub.com/submit-errata`, selecting your book, clicking on the **Errata Submission Form** link, and entering the details of your errata. Once your errata are verified, your submission will be accepted and the errata will be uploaded to our website or added to any list of existing errata under the Errata section of that title.

To view the previously submitted errata, go to `https://www.packtpub.com/books/content/support` and enter the name of the book in the search field. The required information will appear under the **Errata** section.

Piracy

Piracy of copyrighted material on the Internet is an ongoing problem across all media. At Packt, we take the protection of our copyright and licenses very seriously. If you come across any illegal copies of our works in any form on the Internet, please provide us with the location address or website name immediately so that we can pursue a remedy.

Please contact us at copyright@packtpub.com with a link to the suspected pirated material.

We appreciate your help in protecting our authors and our ability to bring you valuable content.

Questions

If you have a problem with any aspect of this book, you can contact us at questions@packtpub.com, and we will do our best to address the problem.

1
Getting Started with Gephi

In this chapter, we will cover the following recipes:

- ▶ Installing Gephi
- ▶ Troubleshooting the Gephi installation
- ▶ Exploring Gephi's graphical user interface
- ▶ The basics of working in the Overview mode
- ▶ The basics of working in the Data Laboratory mode
- ▶ The basics of working in the Preview mode

Introduction

Gephi is an interactive graph and network analysis and visualization tool that allows its users to study the properties of graphs and networks in detail, without having to write any code. Gephi supports almost all types of graphical networks including complex networks, hierarchical networks, dynamic networks, and temporal networks. Gephi has a lot of ready-to-use features that allow users to create stunning and informative visualizations. Graph analysis is one of the preliminary steps in the process of studying graphical systems and Gephi aids in that process by freeing the user from requiring knowledge of programming.

Gephi was developed in Java and, hence, is a cross-platform application, which means it can work on Windows, Linux, and Mac OS X. This chapter will take you through the step-by-step process involved in installing Gephi on different platforms. We will also discuss troubleshooting that might be required during the installation process.

This chapter also gives you an overview of Gephi's **graphical user interface** (**GUI**) and a basic understanding of various modes available in it.

Installing Gephi

This recipe discusses the minimum system configurations required in order to install Gephi and the installation process for different platforms.

Getting ready

Gephi, being a network analysis and visualization tool, requires a compatible graphics card to be installed on your system. It uses a built-in OpenGL engine for fast processing when dealing with very large networks. Hence, it requires OpenGL 1.2 installed on your system. Gephi also requires Java 6 or later. Make sure you have these two programs installed on your system before you go ahead with the installation process.

How to do it...

Follow these steps to install Gephi on Windows:

1. Download the Gephi installer from the official website: `https://gephi.github.io`.
2. Run the installer and hit **Next**:

3. Accept the license agreement and hit **Next**:

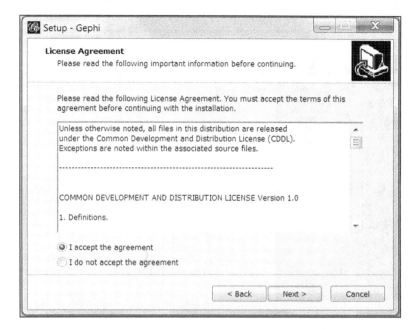

4. Keep clicking on the **Next** button until you reach the following screen. Click on **Install**:

5. The installation will begin and once it's over, the following screen is displayed. Hit **Finish** to complete the setup:

To install Gephi on a Linux machine, follow these steps:

1. From Gephi's official website (`https://gephi.github.io`), download the TAR installer file.

2. Once the download has finished, untar the file and run Gephi by executing `./bin/gephi`.

3. Ubuntu users can make use of the daily build, available from Launchpad. First, run `sudo apt-add-repository ppa:rockclimb/gephi-daily` to your software sources.

4. Then, run `sudo apt-get update`, followed by `sudo apt-get install gephi`, in order to install Gephi on your box.

The installation process for Mac OS X is pretty simple and straightforward:

1. From Gephi's official website (`https://gephi.github.io`), download the installer.

2. Once the download has finished, double-click on the DMG file to run the installer.

3. Once the installation has finished, a new window will open. Double-click on the Gephi icon to run the application.

There's more...

As mentioned earlier, Gephi is a cross-platform tool and works on many more platforms than the ones described in this recipe. If the graphs on which you are going to work are very large, then a 64-bit version of Gephi will have to be installed if you are also using a 64-bit machine.

One might face some issues while installing Gephi, most of which are related to memory management in Java. To learn about fixing some of the most common ones, refer to the next recipe, *Troubleshooting the Gephi installation*.

See also

- ▶ https://gephi.github.io/users/install/ for documentation on installing Gephi on other platforms

Troubleshooting the Gephi installation

While installing Gephi, there are a couple of problems that users encounter quite frequently, most of which are related to memory-specific requirements. Some of these issues are discussed in this recipe, along with the fixes that one can employ to resolve them.

How to do it...

If you encounter any memory- or JVM-related issues, try following these steps to check if the issue can be resolved:

1. If you are using Java 8, try downgrading to Java 7 and check whether the problem is resolved.

2. If you are using the latest version of Gephi, uninstall it and install an older version. If these two fixes do not resolve the problem, then you might need to do operating system-specific fixes, as listed in the following points:

    ❑ For Windows systems, go to the Gephi folder in `Program Files` in `C:\` and then go to the `etc` folder. Open the `gephi.conf` file in Notepad. In the default options, change the `-J-Xmx512m` value to `-J-Xmx1024m`. This changes the maximum heap size allocated to Java to 1,024 MB. If you are using a 64-bit machine, the `gephi.conf` file will be located in the Gephi folder in `C:\Program Files(x86)`.

- For Linux systems, go to the `etc` folder in the Gephi application directory and open the `gephi.conf` file. Change the `-J-Xmx512m` value to `-J-Xmx1024m` to change the maximum heap size allocated to Java to 1,024 MB.

- For Mac OS X systems, go to **Show Package Contents** by right-clicking on the Gephi icon in the `Applications` folder. Inside the `Contents` folder, go to the `Resources/Gephi/etc` folder. Open the `gephi.conf` file and, in the default options, change the `-J-Xmx512m` value to `-J-Xmx1024m`. This changes the maximum heap size allocated to Java to 1,024 MB.

 Unable to save the modified `gephi.conf` file? Open your text editor in administrator mode, navigate to the folder where the file is located, open the file, and then make the changes. Finally, hit **Save** to save the changes.

How it works...

Java Virtual Machine (**JVM**) is an abstract computing machine, otherwise known as a virtual machine. A virtual machine emulates a part of the computing system. JVM executes a Java program compiled into Java bytecode. Since Gephi runs on JVM, its functioning depends on the memory allocated in the system for Java. If very little memory has been allocated to Java, it won't have enough resources to load all the data and, hence, the application won't start. On the other hand, if too much memory has been allocated to Java, then the system won't let Java start and will throw the "JVM Creation failed" message.

See also

- `http://docs.oracle.com/cd/E13150_01/jrockit_jvm/jrockit/geninfo/diagnos/garbage_collect.html` to understand more about memory management in Java

Exploring Gephi's graphical user interface

Gephi offers a very user-friendly GUI to users, making it very easy for novices to explore and manipulate networks with just a few clicks of the mouse.

Getting ready

This recipe describes some of the main GUI components of Gephi and gives the user an overview of what can be achieved. For this, you need to make sure that you have Gephi installed on your system.

How to do it...

To explore Gephi's GUI, perform the following steps:

1. Run Gephi on your system. You'll be welcomed with a small screen, as shown in the following screenshot, that asks you to choose between loading a preexisting sample and creating a new project:

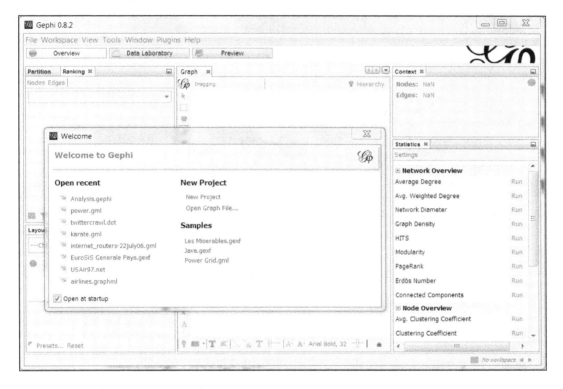

2. If it is your first time with Gephi, click on **Les Miserables.gexf** to load an existing network that was designed using the property of concurrence between the characters from the famous novel *Les Misérables* by Victor Hugo.

3. You will notice a prompt titled **Import report** asking you to set the properties for the graph. Just leave the pre-specified selections as they are and hit **OK** to load the graph.

The following screenshot shows the first screen you will see once the graph has been loaded:

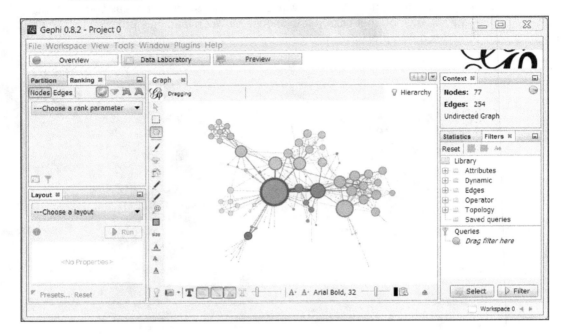

4. To reposition the graph on the screen, place the mouse pointer somewhere on the **Graph** panel situated in the center of the screen and, while holding the right mouse button, move around to the location you want the graph to be centered at.

There are two ways in which the graph can be zoomed-in and zoomed-out:

▶ Rotate the mouse wheel upwards to zoom-in and downwards to zoom-out with the mouse pointer located somewhere in the graph window.

▶ Click on the upward-pointing arrow on the right-bottom corner of the graph window to expand the **Settings** panel. Click on the **Global** tab and use the zoom slider to zoom-in and zoom-out on the graph.

How it works...

Gephi has a pretty simple and user-friendly GUI. The upper-left corner on the screen has three tabs namely **Overview**, **Data Laboratory**, and **Preview**, which represent the three modes present in Gephi for network manipulation. You can customize which panels appear in the application by selecting/deselecting specific panels from the **Window** option in the menu bar.

The basics of working in the Overview mode

This recipe will take you through the basics of various functionalities available in Gephi's Overview mode.

Getting ready

Run Gephi and load a preexisting network. The first screen that you see is the **Overview** mode, which is otherwise called the Graph Manipulation mode in Gephi. If you already have Gephi running, clicking on the **Overview** tab in the upper-left corner of the screen will take you to this mode.

How to do it...

When in **Overview** mode, you'll able to perform a wide variety of manipulations on the graphs. These are categorized under the following subsections in the **Overview** mode, each located in a different part of the **Overview** screen:

- ▶ **Partition**: This module lets you partition the graph into smaller components based on various node- and edge-specific properties, which are called **partitioning parameters**. One such example of partitioning parameters provided by Gephi is **Modularity Class**. The following screenshot shows the graph obtained after partitioning the Les Misérables graph on the basis of **Modularity Class** and then recoloring it:

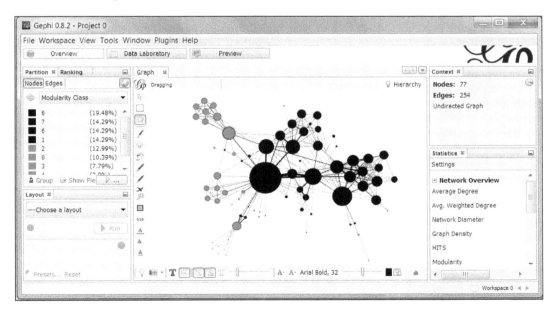

- **Ranking**: This module lets you rank the nodes of the graph based on various criteria such as degree, modularity class, edge weight, and so on. The following screenshot shows the Les Misérables network after its nodes have been ranked, according to their degrees:

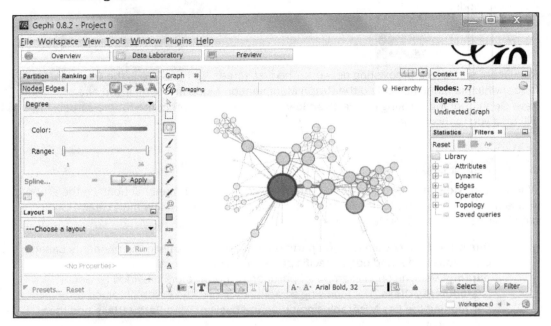

- **Layout**: In this module, one can obtain different visualizations of the same graph by using numerous graph layout algorithms that are provided in Gephi. Some of the most popularly used graph layout algorithms used in Gephi are Force Atlas, Fruchterman Reingold, and Yifan Hu. One such example is shown in the following screenshot, in which the Fruchterman Reingold layout algorithm has been applied to the Les Misérables graph:

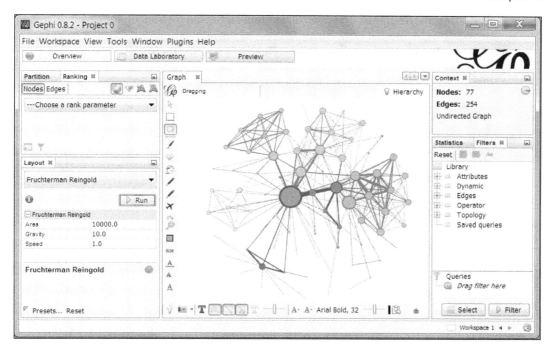

- ▶ **Graph**: In this panel, you'll find a list of basic tools that you can use to perform basic manipulations on the graph such as resizing individual nodes, coloring components of the graph, coloring individual nodes, and modifying node labels. If you do not see this panel on your screen, click on **Window** in the menu bar and select **Graph** from there.

- ▶ **Context**: This part of the **Overview** screen gives information about the basic properties of the graph such as the number of edges, the number of nodes, and the type of graph.

▶ **Statistics**: In this panel, one can run various statistical metrics on the graph to get a deeper insight into the network structure and network properties. In order to get any of the metrics for the graph, simply hit **Run** for a specific metric. This will generate an HTML report, which depicts that metric for the graph. One such report for degree distribution of the Les Misérables graph is shown in the following screenshot:

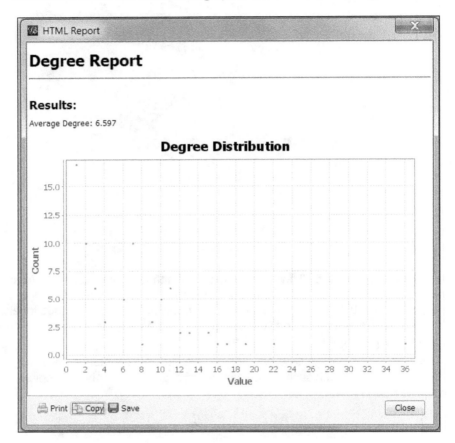

- **Filter**: This part of the **Overview** screen contains various filtering parameters that are based on which of the nodes and edges of the graph could be filtered. One can combine multiple filtering parameters by using the logical operators present in the panel. The following screenshot shows the result of applying the k-core topology filter on the Les Misérables graph with the **k** value being **5**:

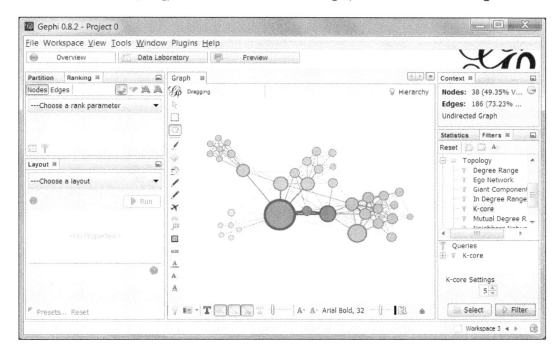

See also

- *Chapter 3, Using Graph Layout Algorithms*, to understand more about various graph layout algorithms present in Gephi.

- *Chapter 4, Working with Partition and Ranking Algorithms*, to understand different ranking and partitioning approaches for graphs.

- *Chapter 5, Running Metrics, Filters, and Timelines*, to know more about statistical metrics and filters for graphs. This chapter also explains concepts related to timeline controls for dynamic graphs.

- http://en.wikipedia.org/wiki/Degeneracy_(graph_theory)#k-Cores to understand more about k-core topology in graphs

The basics of working in the Data Laboratory mode

This recipe introduces the **Data Laboratory** mode in Gephi, which allows users to manipulate network information represented in a tabular format.

Getting ready

Run Gephi and load the Les Misérables network.

How to do it...

To understand the basics of the **Data Laboratory** mode in Gephi, follow these steps:

1. Click on the **Data Laboratory** tab, which is placed next to the **Overview** tab in the upper-left corner of the screen. You will see that all the information about the network can now be seen as a data table, as shown in the following screenshot, with columns or attributes such as nodes, node ID, and node label:

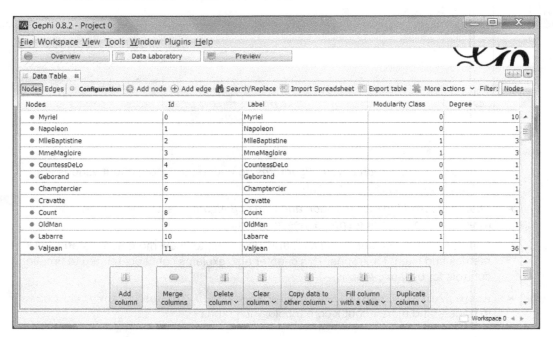

2. Click on the **Edges** button, which is located right below the **Overview** tab in the upper-left corner of the view, to switch to viewing edge-specific details of the network. This is depicted in the following screenshot:

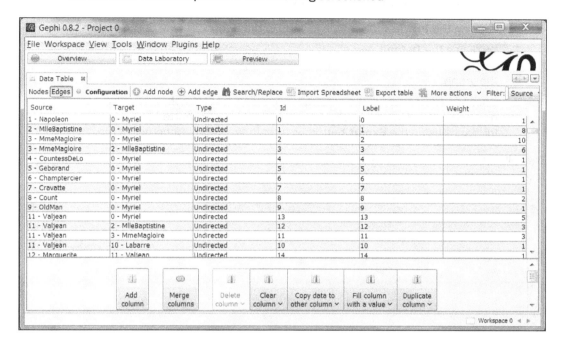

3. In this mode, you can carry out tasks such as adding new columns, deleting columns, merging columns, and importing and exporting spreadsheets. Note that the columns here refer to the attributes of the edges.

See also

▶ Chapter 6, *Working in the Data Laboratory Mode*, to learn more about the functionalities offered in the **Data Laboratory** mode in Gephi

The basics of working in the Preview mode

This recipe introduces the fundamentals of the **Preview** mode in Gephi. This mode lets you alter the way the final network will look. One can then export a snapshot of it into a required format such as a PDF or a PNG image file.

Getting ready

Run Gephi and load the Les Misérables network.

How to do it...

The following steps take you through the **Preview** mode in Gephi:

1. Click on the **Preview** tab that is located in the upper-left side of the screen.

2. You'll see the **Default** preset automatically selected. You can change the various properties of the graph in the panel according to how you would like to visualize the graph.

3. Once you have entered the required properties, hit **Refresh**. This will generate the network in the default visualization; it could then be exported to the required file format by using the **Export** button present in the lower-left corner of the screen.

The following screenshot shows the Les Misérables network when visualized in the **Default** preview mode:

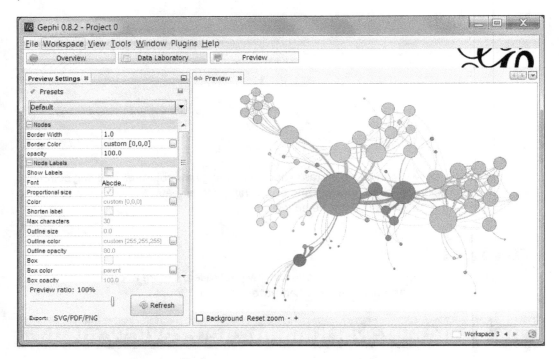

See also

► Chapter 7, *Getting Graphs and Networks Ready for Preview*, to know more about the **Preview** mode in Gephi

2
Basic Graph Manipulations

In this chapter, we will cover the following recipes:

- ▸ Generating a random graph
- ▸ Selecting and highlighting nodes in the graph
- ▸ Coloring and sizing nodes in the graph
- ▸ Adding nodes and edges to the graph
- ▸ Editing node attributes in the graph
- ▸ Finding out the shortest path in the graph
- ▸ Setting the edge and label properties
- ▸ Setting basic properties for graphs, nodes, edges, and labels
- ▸ Changing the background color of the graph
- ▸ Generating a heat map for the graph
- ▸ Showing convex hulls for a graph where a hierarchy exists
- ▸ Showing/hiding various parts of the graph
- ▸ Reverting changes in the graph to the original presets
- ▸ Creating a PNG file directly from the graph window in user-specified sizes

Introduction

This chapter teaches you how to perform basic graph manipulations in Gephi. Manipulations such as modifying network structure by adding or deleting nodes and edges, editing node attributes, and applying filters on the network can be very easily done in Gephi, owing to its user-friendly interface.

Generating a random graph

A random graph with *n* nodes is a graph generated by starting with *n* nodes with no edges existing between any pair of nodes, and then randomly adding edges between nodes in a probabilistic fashion. This recipe describes the process of generating one such random graph in Gephi.

How to do it...

The steps to generate a random graph with *n* nodes are as follows:

1. Click on **File** in the menu bar.
2. Choose **Generate** in the drop-down menu.
3. Choose **Random Graph** in the extended menu. You'll see the following pop-up window on the screen:

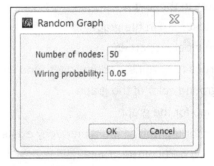

4. In the **Number of nodes** textbox, enter the number of nodes (n) that you want the random graph to have.
5. In the **Wiring probability** textbox, enter the probability (p), according to where you would like the edges to be added between different pairs of nodes in the random graph. A low wiring probability will result in a regular lattice. On the other hand, the higher the wiring probability is, the more random the resulting graph will be.
6. Click on **OK**. This generates the random graph with *n* nodes and probability p; each edge is added between every pair of nodes in the graph.

The following screenshot shows a random graph with 50 nodes and wiring probability of 0.05:

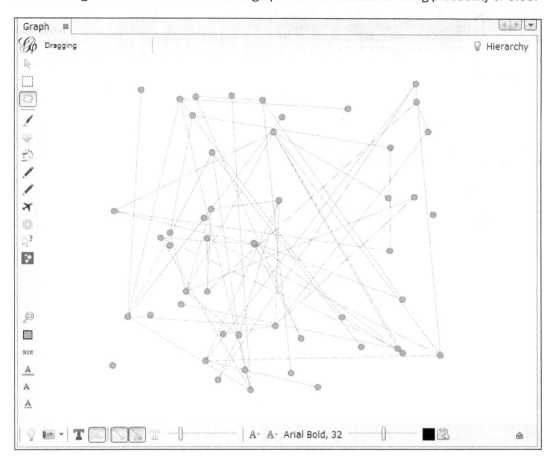

There's more...

Refer to *On random graphs* by Paul Erdős and Alfréd Rényi, published in 1959, at `http://www.renyi.hu/~p_erdos/1959-11.pdf`. It is the first paper that appeared on random graphs.

Research and read more about random graphs on the Web to understand in detail about their properties and the real-world scenarios in which these graphs occur. Random graphs occur very frequently in our day-to-day lives. In fact, most of the networks around us can be modeled as random graphs. Some examples include wireless networks; social networks; the Internet; and food, web, and biological collaboration networks. Refer to the paper titled *Random Graphs as Models of Networks* by M.E.J. Newman at `http://www.santafe.edu/media/workingpapers/02-02-005.pdf` for a detailed insight into this topic.

Selecting and highlighting nodes in the graph

In this recipe, you will learn how to selectively pick up specific nodes in the graph and how to highlight them for closer study.

How to do it...

There are two different ways in which nodes can be selected in Gephi: direct selection and rectangle selection. Here's how we go about selecting the nodes in Gephi:

1. To directly select a node and view its neighbors/adjacent nodes, click on the little arrow button towards the upper-left corner of the **Graph** panel, as shown in the following screenshot:

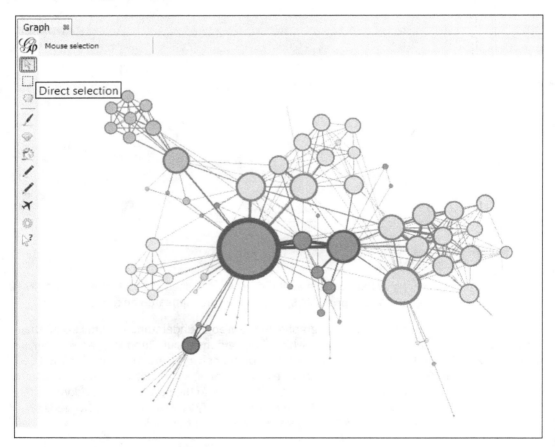

2. Navigate to the node that you want to highlight and place the arrow on the node. You will notice that the node on which you have placed the arrow, along with the neighboring nodes and the edges connecting them, has become highlighted while the other nodes are dimmed. This is shown in the following screenshot:

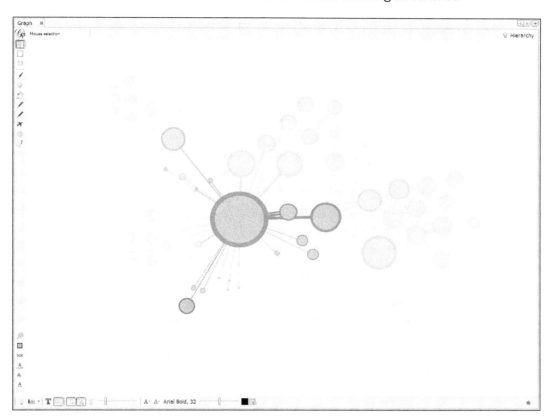

3. To select a node using rectangle selection, click on the button with the little rectangle depicted on it towards the upper-left corner of the **Graph** panel:

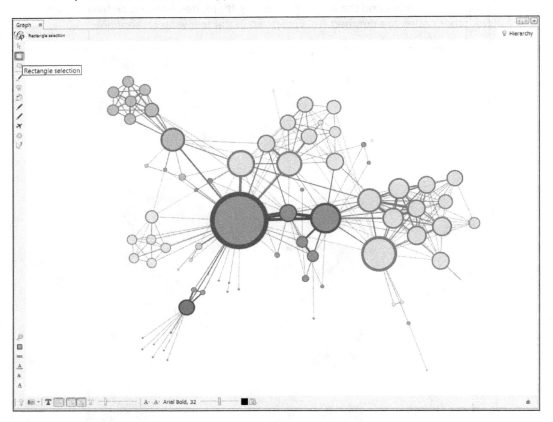

4. Place the mouse pointer somewhere in the vicinity of the node to be selected. Click and hold the left-mouse button while dragging the mouse in order to form a rectangle:

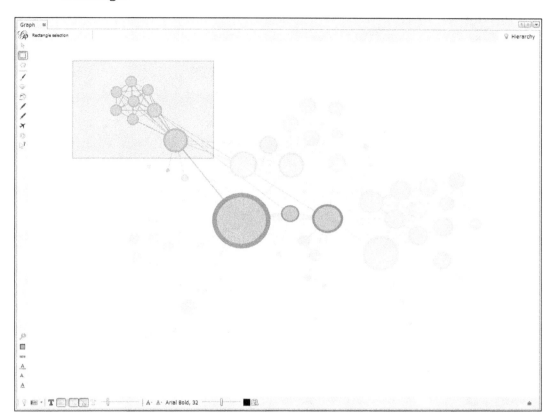

5. Release the mouse when the required area has been selected. This will highlight the nodes in the selected area and all of their neighbors along with the edges connecting them:

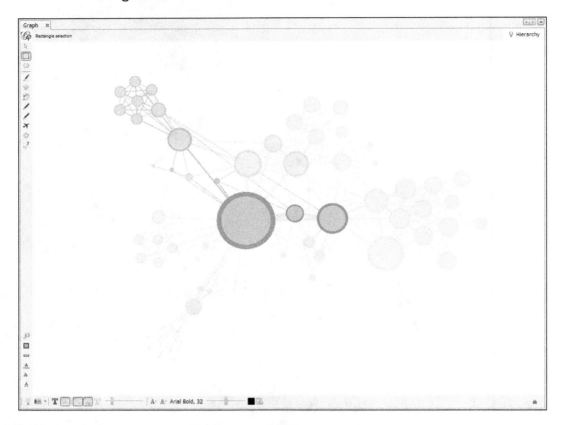

6. To select a single node in the rectangle selection mode, simply click on the node that has to be highlighted.

7. If you want to select multiple nodes one-by-one, hold down the *Ctrl* key and click on the nodes that you want to highlight. Release the *Ctrl* key once you are done selecting all the required nodes. On Mac systems, instead of the *Ctrl* key, use the *Cmd* key.

How it works...

In this recipe, you will have learned the two ways of selecting and highlighting the nodes of a network in Gephi. The selection and highlighting of specific nodes in the graph helps one to focus on a particular section of the graph and study its properties in a more detailed manner. However, there's a significant difference between the two approaches. Using the direct-selection approach, a single node can be selected and highlighted, along with its neighbors. On the other hand, using the rectangle selection approach, a bunch of nodes can be selected at the same time simultaneously.

The other difference lies in the persistence of the selection. Direct selection only results in selecting and highlighting a node temporarily; that is, the nodes remain highlighted as long as the pointer is placed over the node. As soon as the pointer is moved away, the highlighting disappears. On the other hand, when a rectangle selection is used, the highlighting remains persistent and is removed only when the user intends to do so by clicking on a different area of the screen.

There's more...

If you want to select only a specific node(s) without the neighbors and connecting edges, click on the upward-pointing arrow placed at the lower-right corner of the **Graph** panel to expand the toolbar. Click on the **Global** tab in the toolbar and uncheck the checkbox next to **Autoselect neighbor**. This is shown in the following screenshot:

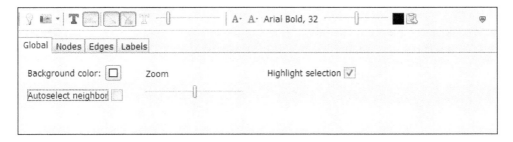

To remove the selection, simply click on any area of the window other than the selected area.

Coloring and sizing nodes in the graph

Sometimes, while studying the properties of a network, we need to manually alter some of the attributes of the nodes, such as their color and size. This section explains how to achieve this.

How to do it...

The following steps illustrate how to resize individual nodes in the graph:

1. Click on the diamond-shaped button in the toolbar placed on the left-hand side of the **Graph** panel. The mouse pointer, when placed over it, should show a descriptor textbox that reads **Sizer**.

2. Click on the node that you want to resize and while holding the mouse button, drag the mouse vertically upwards or downwards. Dragging the mouse upwards will increase the size of the node and dragging it downwards will decrease the size.

3. If the resized node now hides a part of the graph that you otherwise want to continue looking at, click on the button in the same toolbar with a hand symbol on it. Placing the mouse pointer on this button should show up a descriptor textbox that reads **Drag**.

4. Now click on the node that you want to reposition and while holding down the mouse button, drag it to the desired location.

If you want to recolor individual nodes in the graph, follow these steps:

1. Click on the paintbrush-shaped button in the left-hand side toolbar of the **Graph** panel. The mouse pointer when placed over it should display a descriptor box that reads **Painter**.

2. Click on the **Color** box and select the desired color that you would like to assign to the node(s). Click on **OK**:

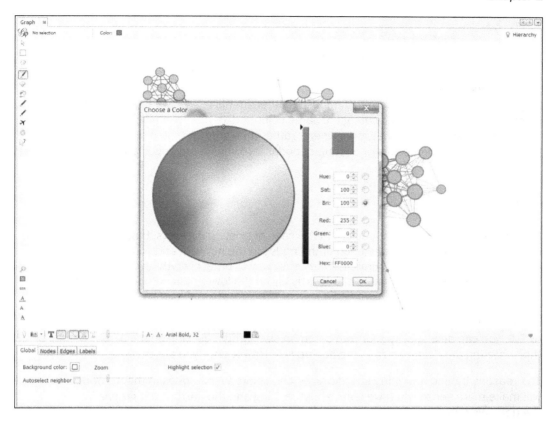

3. Now click on the node to which you would like to assign the selected color. Perform multiple clicks to get the desired shade of color.

4. To assign the same color to multiple nodes, select the nodes to be colored using rectangle selection, as described in the *Selecting and highlighting nodes in the graph* recipe and then click on any one of the nodes. This assigns the same color to all the nodes that were selected.

How it works...

Size and color are two of the most important attributes of the nodes in a graph. They help us study the network in a much more effective manner. Size and color, on one hand, help in differentiating one node or a group of nodes from one another and, on the other hand, help in visualizing related groups of nodes together. In this recipe, we have manually created communities of nodes by allocating the same color and/or size to similar nodes. This will aid us in studying complex networks where, rather than studying the property of each single node, studying the properties of a group of nodes becomes more important to understand the network better.

There's more...

Sometimes it may happen, especially in the case of existing datasets, that you do not like the size and the colors allocated to each of the nodes in the network and you may want to start with the allocation from the scratch. A very easy way to bring everything down to a common level is to use the reset toolbar, placed vertically on the lower-right corner of the **Graph** panel.

The button with a gray rectangle on it will help reset all the colors and gray out all the nodes in the graph. The button with **Size** written on top of it will resize all the nodes to **1.0**. Instead of **1.0**, you can also specify your own common size by right-clicking on the **Size** button and then entering the desired size in the pop-up box. Hit **OK** once you are done.

The reasons behind assigning specific sizes and colors to the nodes in the network will make more sense when we learn about ranking and the partitioning graphs in the upcoming chapters.

Adding nodes and edges to the graph

Gephi allows its users to alter graphs on-the-fly by offering capabilities such as adding node(s) or edge(s) to the graph in just one click. This recipe discusses how to carry out these actions.

How to do it...

The following steps illustrate the procedure to add a new node to the graph:

1. Click on the pencil-shaped button in the toolbar, placed vertically on the upper-left side of the edge of the **Graph** panel. When a mouse pointer is placed on the icon, it should show a descriptor textbox that reads **Node Pencil**.

2. In the upper-right corner of the **Graph** panel, choose the color that you would like to assign to the node and set the desired size for it.

3. Now, click on the area of the **Graph** panel where you want to place the new node. This will create a new node with the selected color and size, as shown in the following screenshot:

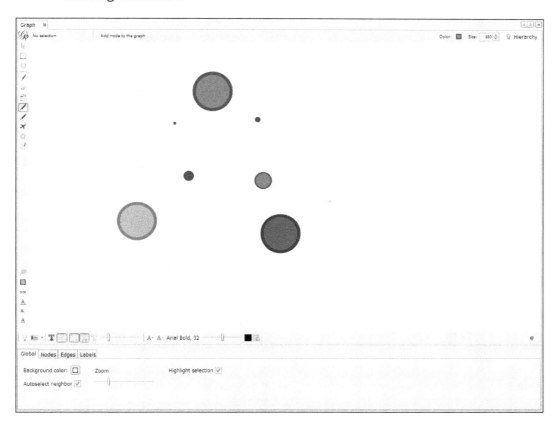

Follow these steps to add a new edge to the graph:

1. Click on the pencil-shaped button in the toolbar that is placed vertically on the upper-left side of the edge of the **Graph** panel. When the mouse pointer is placed on the icon, it should show a descriptor textbox that reads **Edge Pencil**.

2. In the upper-right corner of the **Graph** panel, select the type of edge you want to draw, choose the color that is to be assigned to the new edge, and set a weight for the edge.

3. Now, click on the node that will act as the source node for the edge. Then, click on the target node for the new edge. This draws a new edge of the desired type, color, and weight between the two selected nodes.

How it works...

When we are building up a network from scratch, we need the capability to construct it in a recursive fashion by adding nodes and edges one-by-one. By following this recipe, this task can be carried out in a very user-friendly and efficient manner.

The size of the nodes differentiates one node from the other. For example, in a relationship network, a large node might indicate the prominence and importance of the node over smaller nodes. It might also depict the existence of large number of relationships with other nodes in the network.

The weight of the edge shows how much stronger the relationship is between two nodes in the graph. For example, in a friendship network, a relatively thicker edge between might show that the two people are very close friends whereas a relatively thinner edge might show that the two people are mere acquaintances and do not have a good amount of interaction.

There's more...

If you want to delete a node that you might have added by mistake or that you no longer need, select the node by using any of the selection mechanisms that we discussed in the *Selecting and highlighting nodes in the graph* recipe. Now right-click on the node and select **Delete** to delete the selected node. You can also select the node and press *Ctrl + D* to delete the node.

Editing node attributes in the graph

There might be cases where one would like to change some attributes for specific nodes manually. Gephi offers a very easy way to do so. In this recipe, we will discuss how, with just a single click, you can alter node attributes such as size, position, color, label, and so on.

How to do it...

The following steps illustrate the way to change node attributes such as size, position, color, label, and so on.

1. Click on the button with an arrow and a question mark on it that is placed last on the toolbar located vertically on the upper-left edge of the **Graph** panel. When a mouse pointer is placed over this button, it should read **Edit-Edit node attributes**.

2. Now click on the node whose attributes you want to alter.

3. This opens a new panel titled **Edit** on the upper-left corner of the screen.

4. Click on the attribute(s) that you want to modify and assign new value(s).

There's more...

You cannot modify attributes such as ID, since those are unique values.

Finding out the shortest path in the graph

Finding the shortest path in a graph is one of the problems that is widely encountered in many different situations across many different domains. This is one of the fundamental problems in graph theory. It has applications in domains such as computer networks, inventory optimization, flow networks, and so on. In this recipe, we will learn how to compute and visualize the shortest path in a graph in Gephi.

Getting ready

Load a pre-existing network in which you would like to find the shortest path, such as Les Misérables, or create one.

How to do it...

To compute and visualize the shortest path for a pair of nodes in a network, follow these steps:

1. Click on the button with an airplane symbol on it in the vertically placed left-hand side toolbar in the **Graph** panel. When placed over this button, the mouse pointer displays a descriptor textbox that reads **Shortest Path**.

2. Now click on the source node in the graph from where the path should start.

3. Select the target node to which you want to draw the shortest path in the graph.

   This highlights a path comprising of certain nodes and edges in the graph. This is the shortest path between the source and target nodes selected.

4. If you want to set a different color for the shortest path, choose the desired color in the upper-right corner of the **Graph** panel before selecting the source and target nodes.

The following screenshot shows the shortest path between the nodes labeled **Valjean** and **Grantaire** in the Les Misérables network:

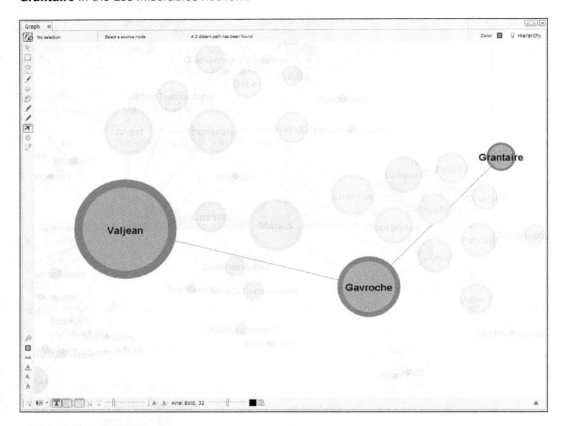

How it works...

Upon selecting the source node and target node between which the shortest path has to be computed, Gephi runs specific algorithms for computing the shortest path.

For computing the shortest path in undirected graphical networks, a very popular algorithm called **Djikstra's algorithm** is used. Undirected graphical networks are networks where all the edges are bidirectional. For directed graphical networks, Gephi makes use of the Bellman-Ford algorithm. Directed graphical networks have unidirectional edges.

There's more...

To remove the highlighted shortest path and compute the shortest path between a different pair of nodes, simply click on any area of the window.

See also

▸ Djikstra's algorithm and the Bellman-Ford algorithm to understand more about how the shortest paths are computed between two nodes in an undirected or directed graph

▸ *Introduction to Algorithms* by Thomas H. Cormen, Charles E. Leiserson, Ronald L. Rivest, and Clifford Stein for a detailed insight into algorithms

▸ `http://optlab-server.sce.carleton.ca/POAnimations2007/DijkstrasAlgo.html` for an interactive tutorial on Djikstra's algorithm

Setting the edge and label properties

Gephi offers functionality to set up edge and label properties such as edge weight, edge color, label weight, label color, and so on. These functionalities will be discussed in this recipe.

How to do it...

To set edge and label properties explicitly, follow these steps:

1. Load the Les Misérables graph in Gephi.

2. Click on the upward-pointing arrow located at the bottom-right corner of the **Graph** panel to expand a new panel.

3. In that panel, click on the **Labels** tab.

4. Check the checkbox located next to **Nodes**. This will display all the node labels.

5. The first button we will discuss defines how the edges will be colored. This is the button with a line segment and a mini-rainbow-like structure as its icon placed at the fourth position in the horizontal toolbar located towards the bottom of the **Graph** panel. Hovering the mouse pointer on top of it should show a descriptor textbox that reads **Edges have source node color**.

6. When toggled on, this will result in the edges taking the same colors as that of each of their respective source nodes. When toggled off, the edges will all acquire the color gray.

7. The next feature is the slider that is used to set the edge weight scale. This is also located in the same toolbar that is located at the bottom of the **Graph** panel. Move the slider back and forth to change the thickness of the edges proportionally to their current thickness.

8. The next button is the one with a black-colored capitalized **A** as its icon. This is also located in the toolbar at the bottom of the **Graph** panel. When a mouse pointer is hovered on top of this button, the descriptor textbox that appears should read **Size Mode**.

9. Clicking on this button will open a drop-down menu with three options: **Fixed**, **Scaled**, and **Node Size**. You can choose your desired option from this menu.

10. The next button adjacent to the button discussed in the previous point is the button to change the color assignment to the labels. This is the button with a capitalized colored **A** as its icon. When a mouse pointer is kept on this button, the descriptor textbox that appears should read **Color Mode**.

11. Clicking on this button will open a drop-down menu with two options: **Unique** and **Object**. Select your desired option for the label colors.

12. Next to the button mentioned in the previous point is the button for selecting the font type and font size for the labels. Clicking on this button opens up the **Font** dialog box.

13. In the **Font** dialog box, select the desired font type, font style, and font size for the labels. Hit **OK** when done.

14. The second slider on the toolbar is the **Font Size Scale** slider that is used to proportionally scale the font size for the labels in the graph. Move the slider back and forth to change the font size.

15. The next button in the toolbar is the **Default Color** button that is used to set the default color for the labels. Left-click on this button and, while holding the mouse button, navigate to a color in the pop-up color palette to quickly set the default color for the labels in the graph.

16. You can also right-click on the button to open up a color palette window from where you can select the default label color. Hit **OK** when done.

17. The last button in the toolbar is the **Attributes** button, used to set the information that has to be displayed with the labels. The button has a small wrench symbol as its icon. Clicking on this button opens up the **Label Text Settings** dialog box.

18. In the dialog box, choose the attributes that you would like to display as labels. There are three different attributes that can be shown: **Id**, **Label**, and **Modularity Class**. Hit **OK** once done:

Setting basic properties for graphs, nodes, edges, and labels

Apart from what has been discussed so far in this chapter, Gephi also offers some additional options for configuring the basic properties of the graphs, nodes, edges, and labels. This section discusses these functionalities.

Getting ready

Load the Les Misérables graph in Gephi.

How to do it...

In the following steps, we will discuss various functionalities offered by Gephi to set the basic properties of graphs, nodes, edges, and labels.

1. Click on the upward-pointing arrow located in the bottom-right corner of the **Graph** panel to expand a new toolbar with four tabs: **Global**, **Nodes**, **Edges**, and **Labels**.

2. Click on the **Global** tab.

3. Click on the button next to **Background Color** to open a color palette window where you can select the background for the graph.

4. The **Autoselect neighbor** checkbox is used to turn the toggle on and off for selecting the neighbors when a node is selected.

5. Move the **Zoom** slider back and forth to zoom-in and zoom-out of the graph.

6. The **Highlight Selection** checkbox is used to set the toggle on and off for the selection highlighting. Check this checkbox and then click on any node in the graph. This will highlight the node selected and its neighbors. Unchecking this box will remove this functionality and result in the selection not being highlighted.

7. Click on the **Nodes** tab.

8. Click on the **Default Shape** menu and it will open a drop-down menu with three options for the shape of the nodes—**Sphere3D**, **Disk2D**, and **Rectangle**.

The following screenshot shows the Les Misérables graph, in which nodes have been shaped as three-dimensional spheres:

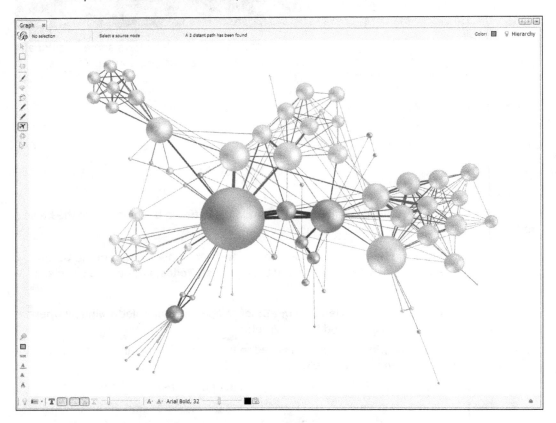

9. The **Show hulls** checkbox is used to toggle on or off, showing the hulls for hierarchical graphs.

10. Click on the **Edges** tab.

11. Uncheck the **Show** checkbox to remove the edges from the graph visualization. Check the box to show the edges again.

12. Using the **Edge default color** button, you can set the default color for the edges. Click on the button to open a color palette window from where you can choose the default color for the edges. Hit **OK** when done. Note that the default color will not be visible until the source node color for edges feature is toggled off. You can achieve this by unchecking the checkbox titled **Source Node Color** that is located beneath the button that we have juts mentioned.

13. The scale slider mimics the **Edge Weight** slider that we discussed in the previous recipe and is used to change the edge weights proportionally to their current weights.

14. Check the **Selection Color** checkbox to set the colors for incoming and outgoing edges for a selected node in a directed network.

15. Click on **Labels**.

16. Check the **Node** and **Edge** checkboxes to enable the node labels and edge labels, respectively.

17. The **Font** selection menu, color, and size mimic the button that we have already discussed in the previous recipe.

Changing the background color of the graph

Gephi, by default, has white as the background color of the graphical area. This can be very easily changed with a single click, which is described in the recipe in this section.

How to do it...

Following are the steps to change the background of the graph/charting area:

1. Locate the button with a bulb icon on it, which is located in the lower-left corner of the **Graph** panel. When the mouse pointer is hovered over it, a descriptor textbox should appear that reads **Background Color**.

2. Left-click the button to alternate the background between black and white.

3. Right-clicking on the button opens up a pop-up window that allows you to select the color for the background. Once chosen, hit **OK**.

Generating a heat map for the graph

Heat maps for graphical networks are a type of visualization to help you understand the distance of one node from other reachable nodes in the graph. Gephi allows us to generate heat maps only for directed networks—that is, only for when all the edges in the network are directed.

Getting ready

While starting Gephi, open the Les Misérables graph in the directed mode by selecting the **Directed** option from the **Graph Type** drop-down list on the screen that comes after you click on **Les Misérables** on the welcome screen. The graph will look something like the one in the following screenshot:

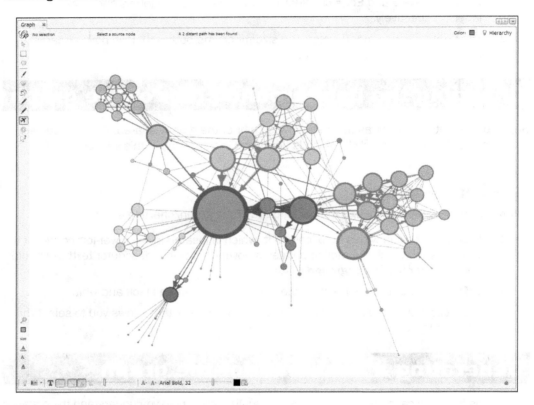

How to do it...

This recipe describes the procedure to generate a heat map on the network with respect to a node in the graph. To do this, perform the following steps:

1. Click on the button on the left-hand side toolbar of the **Graph** panel that has a gear-like icon on it. When placed on it, the mouse pointer shows a descriptor textbox that reads **Heat Map**.

2. Choose the mode for the heat map—**gradient** or **palette**. If you have selected **palette**, select the desired option from the list of available palette options.

3. Now click on the node from which you want to generate the heat map.

How it works...

A heat map is an assignment of an array of colors, usually the same color with different intensities, to nodes in the graph based on a specific property. In this case, the assignment with respect to a source node takes place depending on the distance of each of the reachable nodes from the source node.

For example, if you select a gradient option for the heat map, as shown in the previous screenshot, the darker shades of red will be assigned to the nodes closer to the source node and, as we move away, the lighter shades of red will be assigned.

There's more...

If you want to assign lighter shades to the nearby nodes and darker shades to the distant ones when you have selected the **Palette** mode for the heat map, just uncheck the **Invert Palette** checkbox, which is next to the **Palette** selection menu on the upper side of the **Graph** panel.

Also, if you want to closely study the network without having other distractions, you can reset the node colors to gray by clicking on the gray rectangle located in the vertically-placed toolbar towards the lower-left side of the **Graph** panel that reads **Reset Colors** when a mouse pointer is placed over it. Once the reset has happened, you can generate the heat map for the graph.

Showing convex hulls for a graph where a hierarchy exists

Hierarchical graphs are graphs in which hierarchical relationships exist between the members and hence the graph can be visualized with multiple levels.

A simple example of a hierarchical graph is a family tree. In a family tree, each member is depicted as a node; a relationship (or an edge) exists between two nodes if they are spouses or if they have parent-child relationship between them. This tree can be represented as a hierarchical graph visualized as a vertical-levelled graph with members at each level belonging to the same generation.

A convex hull for a set of points (in our case, nodes of the graph), in simplistic terms, is the minimal closed structure that would enclose all the points inside it. It can be visualized as an enclosure obtained when a rubber band is stretched around the set of points in consideration.

How to do it...

Follow these steps to generate a convex hull for subgraphs in a hierarchical graph:

1. Load the Les Misérables graph into Gephi.

2. Using the rectangle selection, select a portion of the graph.

3. Right-click on any of the nodes selected and click on **Group**, or hit *Ctrl + G*.

4. You will notice that some of the nodes have contracted into a single node. Click on that node and select **Expand** from the pop-up menu or hit *Ctrl + E*.

5. Repeat this process two to three more times to obtain some more groups.

6. Placed at the bottom of the **Graph** panel, you will find a button with a polygon on it; if you place the mouse pointer over it, a descriptor textbox will open up that reads **Show Convex Hull**. Click on that button to turn the **Show Convex Hull** toggle on.

7. You will notice that the groups have been enclosed by a closed polygon. This is the convex hull for that group:

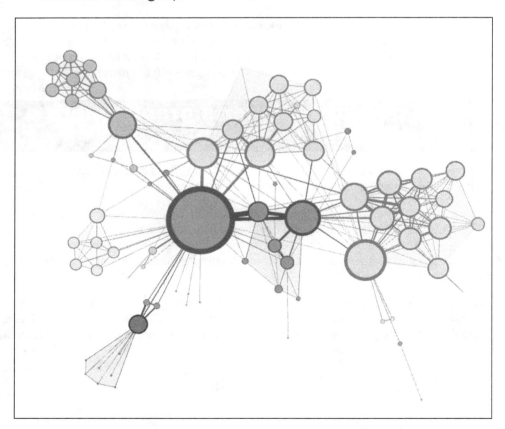

There's more...

Clicking on any of the groups will highlight the convex hull for that group of nodes. Convex hulls help us to visualize the community structure in a network in a better way.

See also

- ▸ http://en.wikipedia.org/wiki/Convex_hull_algorithms for more information on convex hulls. Read about convex hulls and convex hull algorithms to understand the concept covered in this recipe in detail

- ▸ The book titled *Introduction to Algorithms* by Thomas H. Cormen, Charles E. Leiserson, Ronald R. Rivest, and Clifford Stein for a detailed description about convex hulls

Showing/hiding various parts of the graph

The toolbar placed horizontally at the bottom of the **Graph** panel has options to show/hide specific parts of the graph. In this recipe, we will look at those options and understand how they might be useful to us.

How to do it...

Let us explore various options present in the toolbar at the bottom of the **Graph** panel:

- ▸ The first button in the toolbar is the **Show Node Labels** toggle button. This is the button with a capitalized **T** as its icon. Clicking on this button for the first time will display the node labels. Clicking on it again will make the node labels disappear.

- ▸ The second button in the toolbar is the **Show Hulls** toggle button. This is the button with a small enclosed structure as its icon. Clicking on this button will display convex hulls for every subgraph in hierarchical graphs.

- ▸ The third button in the toolbar is the **Show Edges** toggle button. This is the button with a line segment as its icon. Clicking on this button for the first time will make all the edges disappear from the graph. Clicking it again will make the edges reappear.

- ▸ The fourth button on this toolbar is the **Show Edge Labels** toggle button. This is the button with a capitalized **T** in white as its icon. Clicking on this button for the first time will result in the edge labels being displayed, if any. Clicking on it again will result in the edge labels being hidden.

How it works...

The buttons that we learned about earlier are all toggle buttons that can be used to toggle between showing and hiding specific parts of the graph. This helps us choose the amount of detailing on the visualization while studying the graph.

Reverting changes in the graph to the original presets

There are a lot of cases, some of which have already been discussed in the recipes covered so far in this chapter, where one would like to reset the attributes and changes that have been made in the graph. Gephi offers individual reset options for each of the attributes; these will be discussed in this recipe.

Getting ready

The reset buttons are all available in a toolbar that is placed vertically on the lower-left edge of the **Graph** panel. For easy reference and to avoid repetitions, we will call it the **reset toolbar** in this recipe.

How to do it...

Follow these steps to reset the changes in the graph individually, one-by-one.

1. Load the Les Misérables graph in Gephi.

2. To move the graph in the center of the screen and fit it in the window so that the entire graph can be seen as a single entity, click on the button with the magnifying glass icon on it in the reset toolbar. The descriptor textbox that is shown when the mouse pointer hovers on this button should read **Center on Graph**.

3. To reset the node colors back to a base color, which is gray by default, click on the button with a gray rectangle icon on it in the reset toolbar. When the mouse pointer is placed on this button, the descriptor textbox that appears should read **Reset Colors**.

4. If you wish to set the common color to another color rather than gray, right-click on the same button as described in the previous step and choose the color that you would like to assign to all the nodes. Hit **OK**.

5. To reset the size of all the nodes to a common base size, which is 1.0 by default, click on the button with **Size** written on top of it in the reset toolbar.

6. If you want to resize all the nodes to a size different from 1.0, right-click on the **Resize** button described in the previous step and enter the desired size in the pop-up window. Hit **OK** once done.

7. The next set of reset options is related to resets for nodes and edge labels. Before going forward, we will make sure that the labels are displayed in the graph. For this, click on the upward-facing arrow towards the lower-right corner of the **Graph** panel. Click on the **Labels** tab and check the checkboxes for **Node** and **Edge** that correspond to the node labels and edge labels, respectively.

The following screenshot shows the Les Misérables graph with node labels displayed:

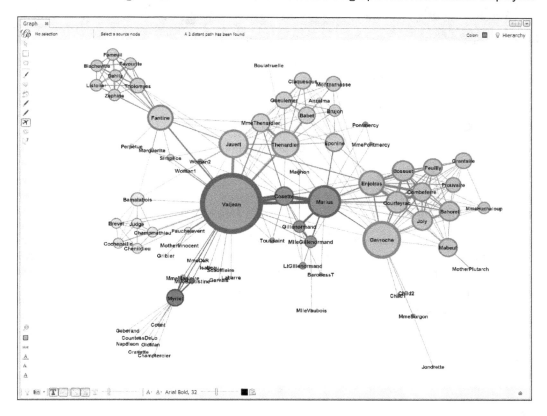

8. The fourth button on the reset toolbar is the **Reset Label Color** button. This is a button with a capitalized **A** with a pink rectangle beneath it. Clicking on this button will set the color of all the labels to the base color.

9. There are certain cases where labels need to be shown only for a part of the graph. These cases will be discussed later in the *Chapter 5*, *Running Metrics, Filters, and Timelines*. To show the labels for the whole graph with a simple click, simply hit the button with a capitalized **A** written on top of it in the reset toolbar. The mouse pointer when hovered on this button should show a descriptor textbox that reads **Reset Label Visible**.

10. To reset the size of all the labels to the default in the graph, click on the button with capitalized **A** and a horizontal arrow displayed on it in the reset toolbar.

See also

▶ *Chapter 5, Running Metrics, Filters, and Timelines*, to learn more about how label-specific attributes can be changed for specific nodes

Creating a PNG file directly from the graph window in user-specified sizes

Quite often, while studying a graphical network, one would like to take snapshots of the graph for future reference. Gephi allows its users to do so in a single mouse click.

Getting ready

Load a preexisting graph or create a new graph in Gephi.

How to do it...

The following steps describe the procedure to take a PNG snapshot of a graph:

1. Locate the button with the camera icon on it placed near the lower-left corner of the **Graph** panel, next to the **Change Background** button. The mouse pointer when placed on this button should display a descriptor textbox that reads **Take screenshot**.

2. Clicking directly on this button will open up the **Save as** dialog box, as shown in the following screenshot, in which you can choose the location where you would like to save the screenshot. Hit **Save** to save the screenshot:

3. If you want to save the screenshot by defining your own dimensions for it, click on the downward arrow placed next to the camera icon. This opens up a drop-down with **Configure** as an option.

4. Click on **Configure**. This opens up **Screenshot settings** dialog box, as shown in the following screenshot:

5. In the **Screenshot settings** dialog box, enter the desired dimensions.

6. Choose the desired antialiasing option from the drop-down menu.

7. Check the **Autosave** checkbox and choose the directory in which you would like to autosave the image file.

8. Hit **OK**.

How it works...

PNG stands for **Portable Network Graphics**. The PNG format enables lossless storage of compressed images. This is one of the most popular image formats used on the Internet, and is mostly used when high-quality images are desired. Gephi only allows screenshots to be saved in the PNG format.

Antialiasing refers to the process of preserving the quality of high-resolution images when they are displayed in lower resolutions. Antialiasing is used to make nodes and edges look smoother.

There's more...

While saving the image, the application may freeze for some time. This is only because the processor is taking time to render the image into the PNG format.

See also

▶ The Wikipedia entry on spatial antialising at `http://en.wikipedia.org/wiki/Spatial_anti-aliasing` for detailed information on spatial antialiasing. Read more about spatial antialiasing in order to understand more about the **Antialiasing** option present in the **Screenshot settings** dialog box.

▶ The article on `http://lunaloca.com/tutorials/antialiasing/`, which gives a good tutorial on understanding antialiasing.

3
Using Graph Layout Algorithms

In this chapter, we will cover the following recipes:

- ▶ Using the Clockwise Rotate layout algorithm
- ▶ Using the Counter-Clockwise layout algorithm
- ▶ Using the Contraction layout algorithm
- ▶ Using the Expansion layout algorithm
- ▶ Using the Force Atlas layout algorithm
- ▶ Using the Force Atlas 2 layout algorithm
- ▶ Using the Fruchterman Reingold layout algorithm
- ▶ Using the Label Adjust layout algorithm
- ▶ Using the Random Layout algorithm
- ▶ Using the Yifan Hu layout algorithm
- ▶ Using the Yifan Hu Proportional layout algorithm
- ▶ Using the Yifan Hu Multilevel layout algorithm

Introduction

While working with graph and network visualizations, one of the most important requirements is to have the ability to visualize the graph with its nodes placed according to some structure across the graphical space. A good tool provides the capability to the users to restructure the network in order to visualize it in a way in which the required parts of the graph are enhanced, similar nodes occupy the same subspace in the graphical space, and all the nodes are clearly distinguishable from each other. This helps give a clear and detailed understanding of the structure of the network. Since networks are not the same and the exploration statement is different in every case, the desired structure in the graphical space might differ. By leveraging the various layouts shown in this chapter, the underlying structure of a network becomes more obvious. This is one of the key advantages of network analysis; going back and forth from one layout to another allows the structure of the network to emerge more quickly.

Various ready-to-use layout algorithms are one of the fortes of Gephi. This chapter covers the default layout algorithms available in Gephi, from the perspective of both conception and implementation.

Using the Clockwise Rotate layout algorithm

In this recipe, we will cover the most basic layout algorithm present in the library of layout algorithms in Gephi. As the name suggests, by using the Clockwise Rotate layout algorithm, one can visualize a network that has been rotated clockwise by 90 degrees (or any other angle for that matter).

Getting ready

Create a new network or load a preexisting one in Gephi. Make sure the **Layout** panel, as shown in the following screenshot, is visible in the Gephi window:

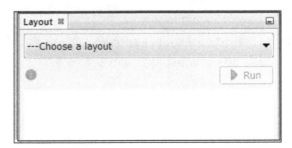

If not, click on the **Window** option in the **Menu** bar and, from the drop-down menu, select **Layout**. The **Layout** panel will appear on the left-hand side of the Gephi application window.

How to do it...

Let's consider the Les Misérables graph. To apply the Clockwise Rotate algorithm on this graph, follow these steps:

1. Load the Les Misérables graph's undirected version in Gephi. Here's how it will appear in the **Graph** window:

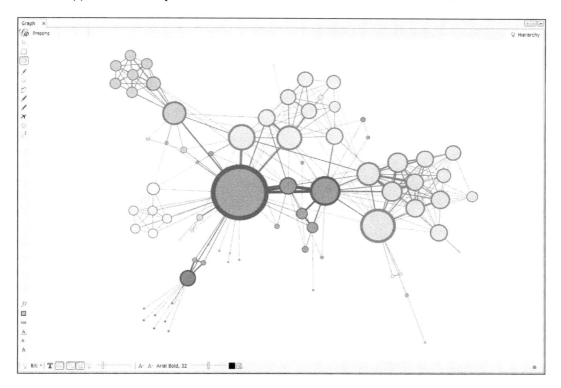

2. In the **Layout** panel, click on the drop-down menu that says **---Choose a layout**:

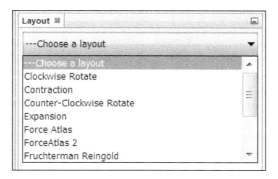

3. From the drop-down menu, select **Clockwise Rotate**. Hovering the mouse pointer over the small round icon with **i** written on it should open a pop-up information box that reads **Clockwise Rotate – Rotate the graph by 90-degrees**.

4. Hit **Run**. The chosen graph, rotated clockwise by 90 degrees, will appear in the **Graph** panel.

The following screenshot shows how the Les Misérables graph will look when rotated clockwise by 90 degrees:

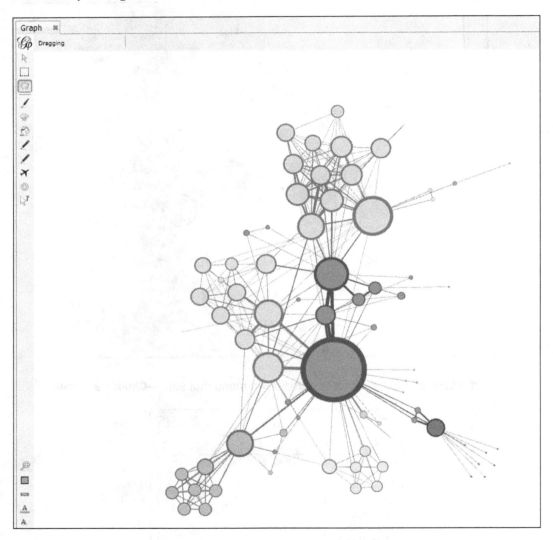

How it works...

We just learned how to rotate a graph by a 90-degrees angle in a clockwise direction. This is one of the geometric transformation algorithms present in Gephi, and may prove to be very helpful in cases where the default layout isn't aesthetically pleasing. The graph, for instance, might be spread more vertically than horizontally and one might want to visualize the same graph spread more horizontally than vertically.

There's more...

If you wish to rotate the graph in a clockwise direction by an angle other than 90 degrees, here's how:

1. In the **properties** box, as shown in the following screenshot, double-click on the textbox that holds the angle figure and enter the angle by which you would like to rotate the graph clockwise:

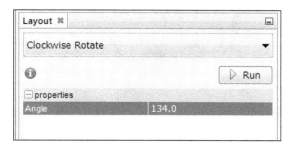

2. Hit **Run**.

The following screenshot shows the Les Misérables graph when rotated clockwise by 134 degrees:

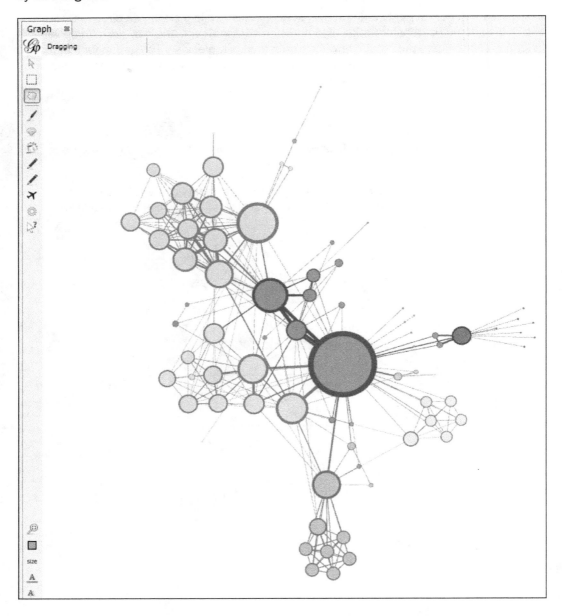

To reset all the values to their defaults, click on the **Reset** button located at the bottom of the left-hand side panel, adjacent to the **Presets** button.

▶   `https://www.khanacademy.org/math/geometry/transformations` for more information on the Geometric Transformation algorithms and the Clockwise Rotate layout algorithm

Using the Counter-Clockwise Rotate layout algorithm

In the previous recipe, you learned how to rotate a graph by specific degrees in a clockwise direction. In this recipe, you will learn about a complementary algorithm to the Clockwise Rotate algorithm. This algorithm is called the Counter-Clockwise Rotate layout algorithm.

The Counter-Clockwise Rotate layout algorithm is used in cases in which rotation of a graph or a network by specific degrees in a counter-clockwise direction is desired.

How to do it...

We will begin with the Les Misérables network and explain how to use the Counter-Clockwise Rotate layout algorithm to obtain a rotated version of the network. The steps remain the same for any other network too. So let's get started.

1.   Load the Les Misérables graph in Gephi.

2.   In the **Layout** panel, click on the drop-down menu that says **---Choose a layout**.

3.   From the drop-down menu, select **Counter-Clockwise Rotate**. Hovering over the small round icon with **i** written on it should open a pop-up information box that reads **Counter-Clockwise Rotate – Rotate the graph by -90 degrees**.

4.   Hit **Run**. The chosen graph, rotated counter-clockwise by 90-degrees, will appear in the **Graph** panel.

The following screenshot shows how the Les Misérables graph will look when rotated in a counter-clockwise direction by 90 degrees:

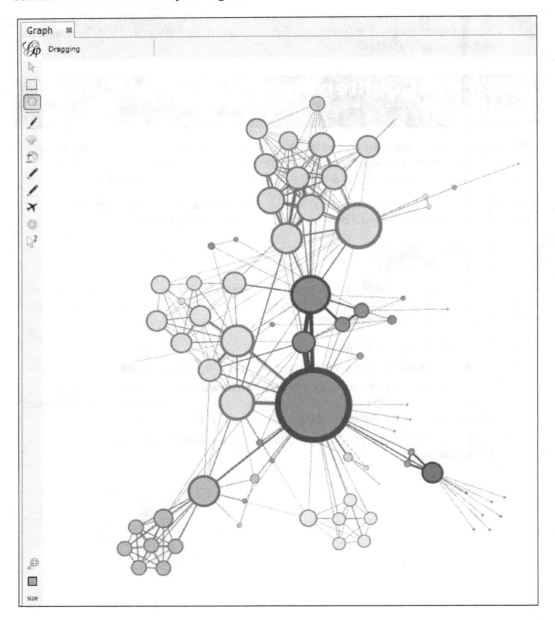

How it works...

You just learned how to rotate a graph or a network by 90 degrees in a counter-clockwise direction. This, similar to the applications described in the *How it works...* section of the *Using the Clockwise Rotate layout algorithm* recipe, comes in handy when you want a snapshot of the network that is more aesthetically pleasing and/or is more symmetrical.

There's more...

If you wish to rotate the graph by an angle other than 90 degrees in a counter-clockwise direction, use the following steps:

1. In the **properties** box, as shown in the following screenshot, double-click on the textbox with **90.0** written in it and enter the angle by which you want to rotate the graph. Hit *Enter* when done.

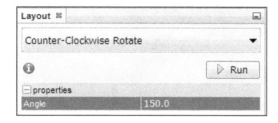

2. Hit **Run**.

The following screenshot shows the Les Misérables network rotated by 150 degrees in a counter-clockwise direction:

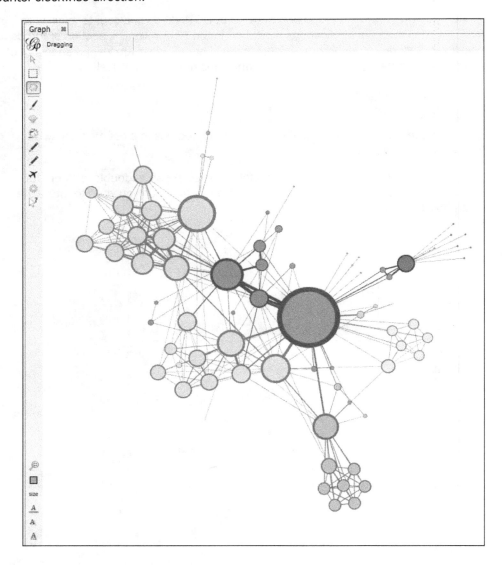

As we mentioned at the beginning of this recipe, the Counter-Clockwise Rotate layout algorithm is complementary to the Clockwise-Rotate layout algorithm; hence, whatever can be accomplished by the Counter-Clockwise Rotate layout algorithm can be accomplished by the Clockwise-Rotate layout algorithm too.

Suppose that you want the network to be rotated by x degrees in a counter-clockwise direction. Simply rotating the same graph in the clockwise direction by (360 - x) degrees would also result in the same final graph.

For example, you want the graph to be rotated in a counter-clockwise direction by 70 degrees, resulting in a different layout for the graph. The same layout can be achieved by rotating the graph in a clockwise direction by (360-70), which is 290 degrees. The following two screenshots proves this.

This screenshot is that of the Les Misérables graph when rotated in a counter-clockwise direction by 70 degrees:

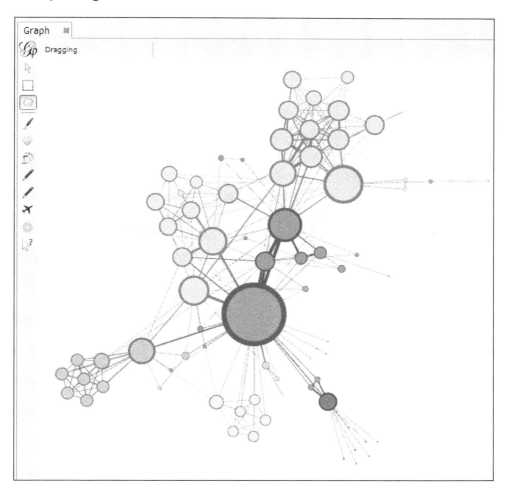

The following screenshot shows the Les Misérables graph when rotated in a clockwise direction by 290 degrees:

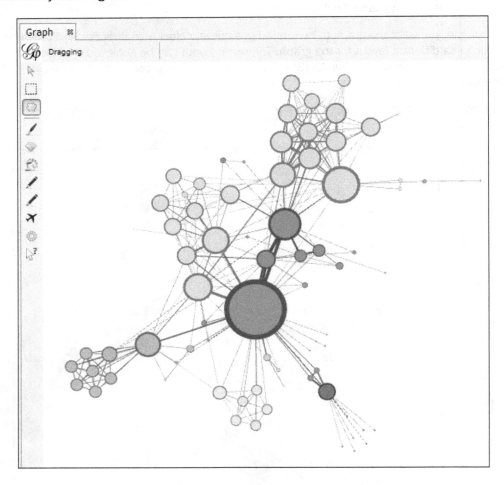

To reset all the values to their defaults, click on the **Reset** button, which is located at the bottom of the left-hand side panel, adjacent to the **Presets** button.

See also

▸ https://www.khanacademy.org/math/geometry/transformations for more information on the Geometric Transformations algorithms

Using the Contraction layout algorithm

There might be instances where nodes are placed too far apart from each other, thereby making the graph appear too sparse. This may lead to difficulty in visualizing the whole network as a single entity. In the simplest case, it may just not be possible to visualize the entire graph on a single window. This is when the Contraction layout algorithm comes into play. Using this algorithm, one can scale down the network and make it appear denser. This recipe takes you through the process of applying this algorithm on a network.

How to do it...

Let us consider Les Misérables network as a way to explain the process of applying the Contraction layout algorithm. The steps remain same for any other network too.

1. Load the Les Misérables graph in Gephi.
2. In the **Layout** panel, click on the drop-down menu that says **---Choose a layout**.
3. From the drop-down menu, select **Contraction**. Hovering over the small round icon with **i** written on it should open a pop-up information box that reads **Contraction – Contracts the layout around its center**.
4. In the properties panel, enter the scale in the **Scale Factor** textbox by which you want the network to be contracted.
5. Hit **Run**. The chosen graph, contracted as desired, will appear in the **Graph** panel.

The following screenshot shows how the Les Misérables graph will look when contracted by 20 percent or by a scale factor of 0.8:

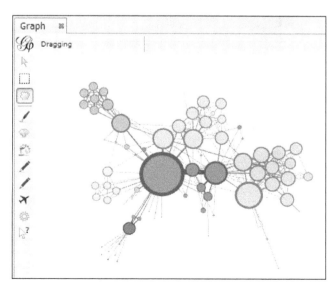

How it works...

The Contraction layout algorithm is yet another Geometric Transformation algorithm present in Gephi. The scale factor defines the ratio of the size of the resulting graph to the original graph. For example, if the scale factor is 0.7, then the resulting graph would shrink by 30 percent of the size of the original graph. What really happens is that every node size remains the same, but the distance between the nodes and the length of the edges is modified according to the scale defined. In this example, the edges will become 70 percent of their original length and the graph gets redrawn.

One might confuse scaling with zooming. There's a subtle difference between the two. During scaling, only the edge lengths change and node sizes remain the same. On the other hand, during zooming out or zooming in, the node sizes change proportionally, as well as edge lengths.

There's more...

There's also the capability to shrink the network according to the user's choice. The following steps describe this process:

1. In the **properties** box, as shown in the following screenshot, enter the scale by which you would like to shrink the network in the **Scale factor** textbox. Hit *Enter* once done.

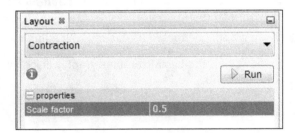

2. Click on **Run**.

The following screenshot shows the Les Misérables network when shrunk by half its size—that is, by a scale factor of 0.5:

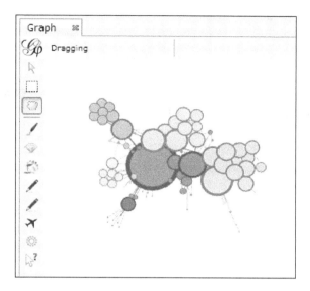

To reset all the values to their defaults, click on the **Reset** button located at the bottom of the left-hand side panel, adjacent to the **Presets** button.

See also

▶ https://www.khanacademy.org/math/geometry/transformations/dilations-scaling/ for more information on graph scaling

Using the Expansion layout algorithm

Sometimes the graph under consideration might be too dense, with the nodes placed very near to each other. Such a graph makes it difficult to visualize the network, thereby hampering its interpretation and understanding. This is where the Expansion layout algorithm comes to the rescue. Using this algorithm, one can scale up the network and make it appear sparser.

How to do it...

Considering the Les Misérables network, the steps to use the Expansion layout algorithm to get an expanded version of the network are as follows. The steps remain the same for any other network, too:

1. Load the Les Misérables graph in Gephi.

2. In the **Layout** panel, click on the drop-down menu that says ---**Choose a layout**.

3. From the drop-down menu, select **Expansion**. Hovering over the small round icon with **i** written on it should open a pop-up information box that reads **Expansion – Expands the layout around its center**.

4. Hit **Run**. The chosen graph, expanded by 1.2 times its original size, will be redrawn in the **Graph** panel.

The following screenshot shows how the Les Misérables graph will look when expanded by 1.2 times when using the Expansion Layout algorithm:

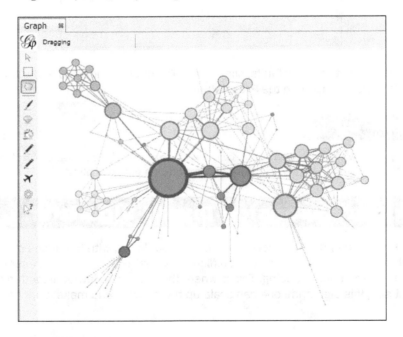

How it works...

The way the Expansion algorithm works is very much the same as the Contraction algorithm. For example, if the user has chosen 2 as the scale, the network will expand to twice its size. This means that the edges will grow to twice their original length, while the node sizes remain the same.

There's more...

Similar to the custom network contraction that was described in the previous recipe, *Using the Contraction layout algorithm*, the user has the ability to view the desired expanded version of the network. To accomplish this, follow these steps:

1. In the **properties** box, as shown in the following screenshot, in the **Scale factor** textbox, enter the scale by which you would like to expand the graph under consideration. Hit *Enter* once done.

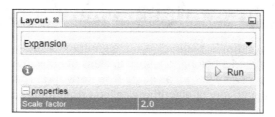

2. Click on **Run**.

The following screenshot shows the Les Misérables network when expanded to twice its original size— that is, by a scale factor of 2.

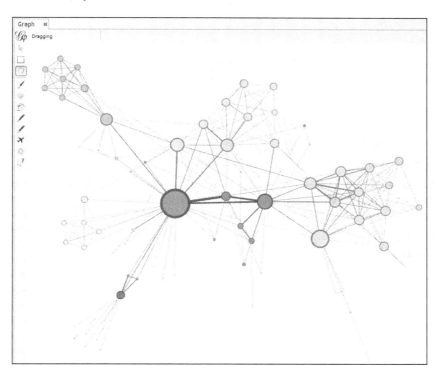

To reset all the values to their defaults, click on the **Reset** button located at the bottom of the left-hand side panel, which is adjacent to the **Presets** button.

See also

There are some good tutorials by Khan Academy on Graph Scaling. Check out the tutorials at `https://www.khanacademy.org/math/geometry/transformations/dilations-scaling/`.

Using the Force Atlas layout algorithm

The Force Atlas layout algorithm is a spatial layout algorithm for real-world networks, such as web networks. Web networks belong to a special class of networks that are known as **small-world networks**, otherwise known as **scale-free networks**. The Force Atlas layout algorithm comes under a category of algorithms called **force-directed algorithms**.

There's usually a trade-off between quality and speed when it comes to graph layout algorithms. Force Atlas emphasizes the former over the latter; that is, the Force Atlas layout algorithm gives more weight to the quality of the layout than the speed with which it has been computed. This is especially true in the case of large networks. In the case of small networks, Force Atlas works just fine.

How to do it...

As with the previous recipes in the chapter, we will begin with the Les Misérables network and explain how to use the Force Atlas layout algorithm for a network. The steps remain the same for any other network too. So let's get started.

1. Load the Les Misérables graph in Gephi.
2. In the **Layout** panel, click on the drop-down menu that says ---**Choose a layout**.
3. From the drop-down menu, select **Force Atlas**.
4. In the **properties** dialog box, enter a very high value, such as `20000.0`, in the **Repulsion strength** textbox. This is to ensure that, at the end of the layout, the nodes do not overlap each other.
5. Hit **Run**. The chosen graph, laid out according to Force Atlas, will appear in the **Graph** panel.

The following screenshot shows how the Les Misérables graph will look when re-visualized by using the Force Atlas layout algorithm with a repulsion strength of 20000.0.

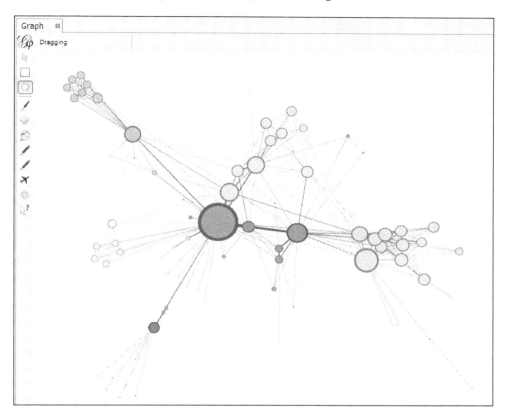

How it works...

As mentioned in the introduction section of this recipe, the Force Atlas layout algorithm belongs to a class of networks known as force-directed algorithms. The force-directed algorithms aim at building up a layout that is aesthetically pleasing with more emphasis on symmetries and non-overlapping nodes. Force-directed algorithms use the properties of the network to produce this kind of layout.

One concept that is made use of in force-directed algorithms is that of hubs and authorities. Hubs in a directed network are nodes with high out-degrees, whereas authorities are nodes with high in-degrees. In the case of undirected networks, treat each edge as a bidirectional edge and then compute in-degrees and out-degrees. Force Atlas enhances the role of authorities while penalizing the hubs in such a way that the authorities get clustered towards the center of the graph, whereas the hubs get laid out towards the periphery.

According to Mathieu Jacomy, one of the initiators of Gephi Project and author of the Force Atlas layout algorithm, Force Atlas's forte lies in its ability to allow the user to study the detailed properties of scale-free networks with the fewest biases possible. The Force Atlas layout algorithm may be slow compared to other layout algorithm, but it produces high-quality results.

There's more...

You may have noticed that there are multiple parameters listed in the **properties** box when the Force Atlas layout algorithm is selected. Each of these parameters helps us to change the settings of the network when recomputed using the Force Atlas layout algorithm. Following are the steps to reconfigure the settings of the Force Atlas layout algorithm and a description of the way they change the final layout:

1. The first attribute that you see in the properties box, as shown in the following screenshot, is **Inertia**. The **Inertia** attribute defines the extent to which the node speed will be conserved at each new pass. That is, it determines how frequently the nodes will change their position in the graphical space with each iteration of the algorithm. For example, an inertia of 0.1 means that the nodes would almost be static in the space whereas an inertia of 50.0 means that the nodes may change their spatial position radically with each pass of the algorithm:

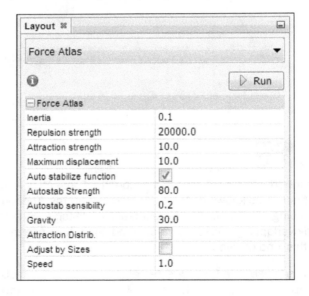

2. The second parameter is **Repulsion Strength**. This parameter defines the force with which each of the nodes will repel other nodes. The following screenshot shows the Les Misérables network when revisualized after running the Force Atlas layout algorithm with a repulsion strength of 5000.0, while other parameters retain their default values:

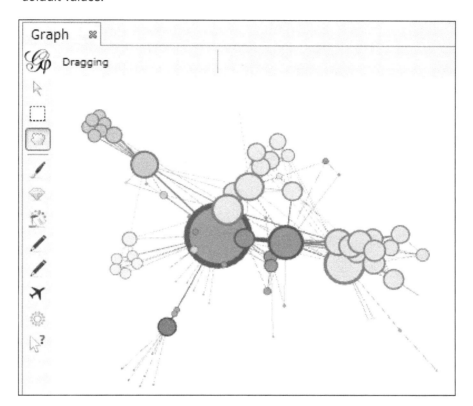

The following screenshot shows the Les Misérables network after the Force Atlas algorithm has been run with a repulsion strength of 20000.0. You can clearly make out how different it is from the previous graph where the repulsion strength was 5000.0:

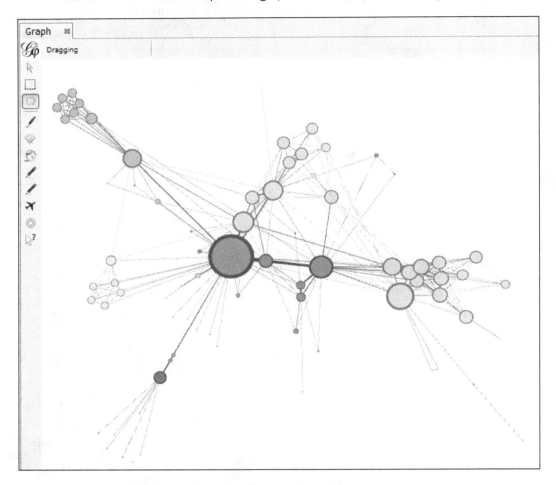

3. The next parameter is **Attraction Strength**, which defines how strongly each pair of connected nodes will attract each other. It is the opposite of repulsion strength with the difference that repulsion strength holds true for all the nodes while attraction strength holds true only for pairs of adjacent nodes.

4. The fourth parameter is **Maximum Displacement**, which is used to put a limit on the amount by which each node can be displaced in the final layout from its initial position.

5. The fifth parameter, **Auto Stabilize Function**, when checked, activates the freezing of the unstable nodes. This helps in stabilizing the network and achieving convergence of the algorithm faster. The selection of this parameter may result in some loss of efficiency.

6. The next parameter, **Autostab Strength**, defines the strength of the **Auto Stabilize** function when it is selected. A high autostab strength will result in infrequent movement of unstable nodes and vice-versa.

7. The seventh parameter is **Autostab Sensibility**, which defines the extent and speed with which the inertia adapts itself during algorithm execution.

8. **Gravity** defines the force with which all nodes are attracted to the center of the graph. This avoids the huge dispersion that might be caused in the case of graphs with disconnected components.

9. One of the defining factors of the Force Atlas algorithm is **Attraction Distribution**. When the checkbox for attraction distribution is checked, the Force Atlas algorithm attempts to centralize the authorities and push the hubs towards the border of the layout. The following screenshot shows how the Les Misérables network will look after the Force Atlas algorithm is run with **Attraction Distribution** checked:

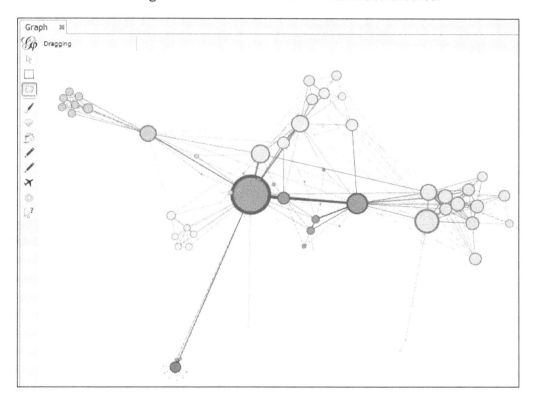

10. The next checkbox is that of **Adjust by Sizes**. When checked, the Force Atlas algorithm attempts to build up a layout with the minimum number of nodes overlapping.

11. The last parameter is the **Speed**. The default value for this parameter is 1. The acceptable values are all values greater than 0. This parameter allows the user to increase the convergence speed of the algorithm at the expense of precision. This means that the Force Atlas algorithm will converge faster but the final layout may not be of good quality.

12. To reset all the values to their defaults, click on the **Reset** button that is located at the bottom of the left-hand side panel, adjacent to the **Presets** button.

There's more...

The Force Atlas layout algorithm is best suited for small-world networks. Small-world networks or scale-free networks mimic the real world. That is, the single-hop direct node-to-node connections in small-world networks are very few but most of the nodes in the graph could be reached from a node in multiple hops. An example could be a friendship network on social networks. If you were to consider the entire network of users of Facebook, it would be a small-world graph since each and every node is connected to relatively few nodes, as compared to the total number of nodes in the entire graph; however, it's possible, in most of cases, to reach any node from a node in a small number of hops.

See also

▸ http://en.wikipedia.org/wiki/Small-world_network for more information about small-world networks

▸ *Chapter 12, Force-Directed Drawing Algorithms*, from *Handbook of Graph Drawing and Visualization* by Stephen G. Kobourov at http://cs.brown.edu/~rt/gdhandbook/chapters/force-directed.pdf to know more about the force-directed layout algorithms

Using the Force Atlas 2 layout algorithm

Force Atlas 2 is another algorithm in the set of force-directed algorithms available in Gephi. Force Atlas 2 attempts to resolve the shortcomings of the Force Atlas algorithm by making a balance between the quality of the final layout and the speed of the computation algorithm. Its performance for large networks is much better when compared to the Force Atlas layout algorithm.

How to do it...

We will begin with the Les Misérables network and explain how to use the Force Atlas 2 layout algorithm on it. The steps remain the same for any other network too. So let's get started.

1. Load the Les Misérables graph in Gephi.

2. In the **Layout** panel, click on the drop-down menu that says **---Choose a layout**.

3. From the drop-down menu, select **Force Atlas 2**.

4. Set **Scaling** to a large value, such as 300.0, otherwise some of the nodes will overlap each other. Hit **Run**.

The following screenshot shows how the Les Misérables graph will look when the Force Atlas 2 layout algorithm is run over it with scaling set to 300.0:

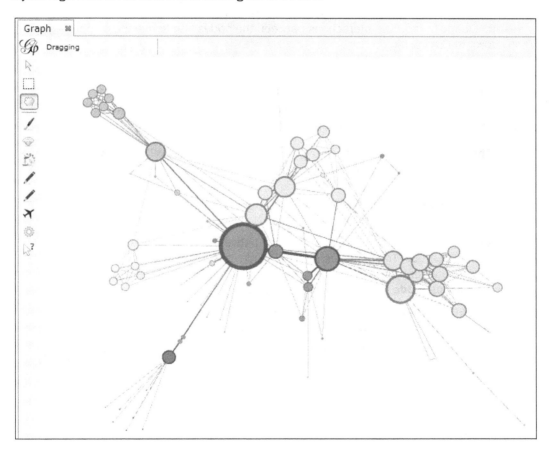

How it works...

The speedup that is achieved in Force Atlas 2 is primarily due to the replacement of direct-sum simulation used in Force Atlas with Barnes-Hut simulation.

The direct-sum simulation tries to analyze the interaction between each entity to every other entity in the system. In a graph, it considers the repulsion force between each node to every other node in the graph and then tries to optimize the overall repulsion. Hence, the direct-sum simulation runs with a complexity of *O(n2)*. On the other hand, Barnes-Hut simulation considers interaction between some entities on an individual one-to-one basis. The remaining entities are clubbed into different partitions and entries in each partition are then replaced by a single representative. In our case, it considers the interaction between directly connected nodes on an individual basis and the rest of the nodes are divided into partitions with a single representative node representing each partition. The algorithm then tries to optimize the repulsion between the considered node and rest of the adjacent nodes or representative nodes. This leads to reduced interactions between the nodes and hence the complexity comes down to *O(n log n)*.

There's more...

The settings for Force Atlas 2 could be modified to get varied layouts for the graph. Here's the explanation of each of the parameters that could be set explicitly:

- **Threads number:** This defines the number of threads that will run in parallel to speed up the execution of the algorithm. This number is limited by the number of cores in the processor.

- **Dissuade Hubs:** This results in the placement of hubs towards the periphery of the network.

- **LinLog mode:** This is the option to switch between the linear-linear mode and linear-log mode. The linear-log (LinLog) mode stands for linear attraction and logarithmic repulsion. By default, it is linear attraction and linear repulsion. The **LinLog mode** results in tighter and closely-knit clusters.

- **Prevent Overlap:** This avoids the overlap of nodes in the final layout.

- **Edge weight influence:** This defines how much weight has to be given to the edge weights. When set to **0**, it will result in computations that do not take into consideration the edge weights. When set to **1**, it gives complete weight to the edge weights and they become crucial to the resulting layout.

- **Scaling:** This defines the closeness of nodes in the resulting graph. A low scaling value will result in a dense graph, whereas a high scaling value will result in a sparse graph.

- **Stronger Gravity:** This defines whether the graph should be drawn with most of the nodes attracted towards the center.

▶ **Gravity**: This defines the strength with which the gravity will be applied to the graph.

▶ **Tolerance**: This is similar to the **Inertia** parameter in the Force Atlas layout algorithm. A lower value results in slow execution of the algorithm but higher precision delivered in the end and vice versa.

▶ **Approximate Repulsion**: This uses Barnes-Hut optimization for optimizing overall repulsion. This results in the computational complexity being reduced from *O(n2)* to *O(n log n)* and hence leads to faster execution.

▶ **Approximation**: This defines the approximation factor for the Barnes-Hut optimization. Refer to the *See also* section of this recipe for more detail on this.

To reset all the values to their defaults, click on the **Reset** button, which is located at the bottom of the left-hand side panel, adjacent to the **Presets** button.

See also

▶ *ForceAtlas2, a Continuous Graph Layout Algorithm for Handy Network Visualization Designed for the Gephi Software* by Mathieu Jacomy, Tommaso Venturini, Sebastien Heymann, and Mathieu Bastian at `http://www.plosone.org/article/ info%3Adoi%2F10.1371%2Fjournal.pone.0098679` for a full explanation of the Force Atlas 2 layout algorithm

▶ `http://arborjs.org/docs/barnes-hut` to understand the Barnes-Hut simulation in detail

Using the Fruchterman Reingold layout algorithm

The Fruchterman Reingold layout algorithm belongs to the class of force-directed algorithms. It is one of the standard algorithms in Gephi and is made use of quite often.

How to do it...

We will begin with the Les Misérables network and explain how to use the Fruchterman Reingold layout algorithm on it. The steps remain the same for any other network, too. So let's get started:

1. Load the Les Misérables graph in Gephi.
2. In the **Layout** panel, click on the drop-down menu that says **---Choose a layout**.
3. From the drop-down menu, select **Fruchterman Reingold**.
4. Hit **Run**. The chosen graph with the new layout appears in the **Graph** panel.

The following screenshot shows how the Les Misérables graph will look with the new layout:

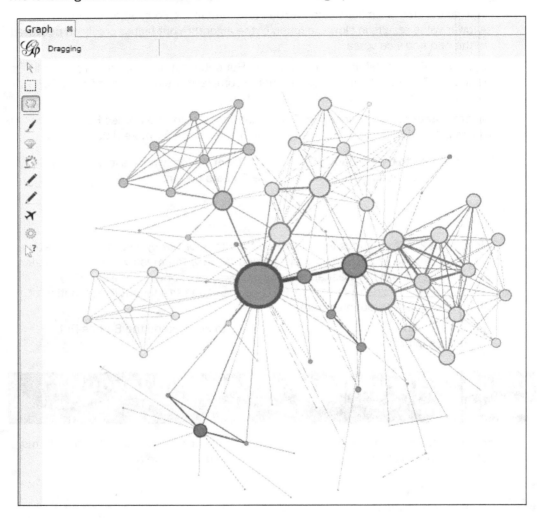

How it works...

In the Fruchterman Reingold layout algorithm, the nodes are assumed to be entities made of steel and the edges are assumed to be springs. The attractive force between the nodes mimics the spring force, whereas the repulsive force between the nodes is analogous to the electrical force. The objective of this algorithm is to minimize the overall energy of the whole system and come up with an optimized layout of the network that satisfies this objective. One important thing to note here is that, unlike the Force Atlas and Force Atlas 2 algorithms, this algorithm does not take into consideration the edge weight to come up with an optimal layout.

There's more...

There are certain parameters in the specification of the Fruchterman Reingold layout algorithm that can be modified to define the final layout of the network. Here's the description of what they are and how they affect the layout of the network:

- ▸ **Area**: This defines the area over which the final graph will be laid out.
- ▸ **Gravity**: This is similar to the concept of gravity discussed under the Force Atlas and Force Atlas 2 algorithms.
- ▸ **Speed**: This defines the speed of convergence of the algorithm. High convergence speed leads to lower precisions and vice versa.

To reset all the values to their defaults, click on the **Reset** button located at the bottom of the left-hand side panel, adjacent to the **Presets** button.

The Fruchterman Reingold layout algorithm works quite well with small and medium graphs but not so well with large graphs, owing to its high computational complexity. So, for large graphs, you might want to explore other layout algorithms or increase the speed in the **Speed** box (but that will lead to lower precision).

See also

- ▸ *Graph drawing by force-directed placement* by Thomas M. J. Fruchterman and Edward M. Reingold at `http://dl.acm.org/citation.cfm?id=137557` and `http://dx.doi.org/10.1002%2Fspe.4380211102` to understand the Fruchterman Reingold layout algorithm in detail

Using the Label Adjust layout algorithm

As the name suggests, the Label Adjust layout algorithm is used to adjust the layout of the labels in the graph. This might come in handy when dealing with dense graphs with a large number of nodes where it becomes necessary to distinguish and be able to read the label distinctly. This algorithm is usually used in conjugation with other algorithms wherein the other algorithm defines the layout of the network and Label Adjust layout algorithm then helps to adjust the labels in the resulting layout.

How to do it...

Using the Les Misérables network, we will explain how to use the Label Adjust layout algorithm to obtain a new version of the network where labels can be read clearly. The steps remain the same for any other network, too. So, let's get started.

1. Load the Les Misérables graph in Gephi.

2. In the **Layout** panel, click on the drop-down menu that says ---**Choose a layout**.

3. From the drop-down menu, select **Label Adjust**. Hit **Run**.

The following screenshot shows how the Les Misérables graph will look when visualized with labels adjusted using the Label Adjust layout algorithm:

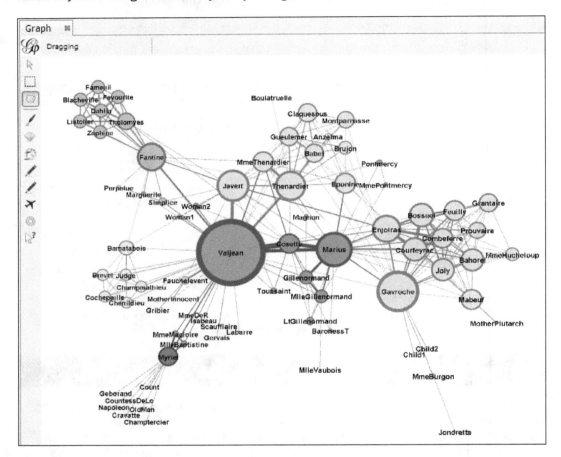

How it works...

The Label Adjust layout algorithm uses the size of the text labels to remodel the network. The label text size acts as the repulsive force to restructure the network. The final layout will be free of any overlapping labels.

There's more...

There's an option to use the node size along with the label text size to define the new layout of the network in this algorithm:

1. In the **properties** box of the Label Adjust algorithm, check the **Include Node Size** box.
2. Hit **Run**.

This means that the larger nodes will repel other nodes more and smaller nodes will repel other nodes less. Here's the screenshot of how the Les Misérables network will look when remodeled by using the Label Adjust layout algorithm with the inclusion of node size:

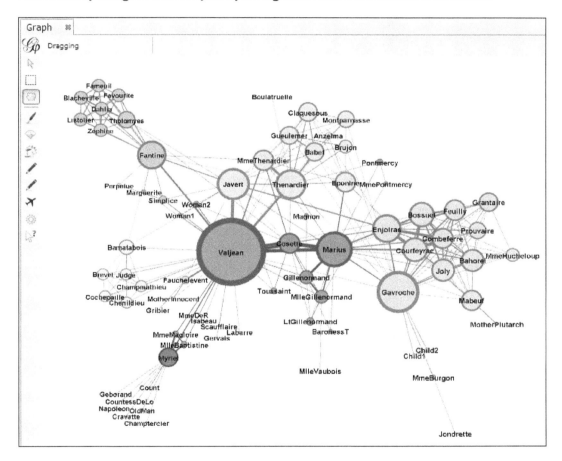

To reset all the values to their defaults, click on the **Reset** button, which is located at the bottom of the left-hand side panel, adjacent to the **Presets** button.

See also

▶ A short video at `http://vimeo.com/2242916` that describes how the Label Adjust layout algorithm works

Using the Random Layout algorithm

The Random Layout algorithm is a Gephi layout that doesn't have a defined purpose to it. It isn't very commonly used either, but sometimes comes in handy when the user isn't looking at anything in particular in the graph and just wants to lay out the nodes in an imaginary rectangular space.

How to do it...

Using the Les Misérables network, we will explain how to use the Random Layout algorithm to obtain a randomized version of the network. The steps remain the same for any other network too. So, let's get started:

1. Load the Les Misérables graph in Gephi.

2. In the **Layout** panel, click on the drop-down menu that says ---**Choose a layout**.

3. From the drop-down menu, select **Random Layout**. Hovering over the small round icon with **i** written on it should open a pop-up information box that reads **Random Layout: A random distribution of the nodes**.

4. In the **properties** box, enter the space size that will define the size of the imaginary rectangular space on which the final graph will be laid out.

5. Hit **Run**. The chosen graph, restructured with random node distribution, will appear in the **Graph** panel.

The following screenshot shows how the Les Misérables graph will look when viewed after applying the Random Layout algorithm with a space size set to `2000.0`:

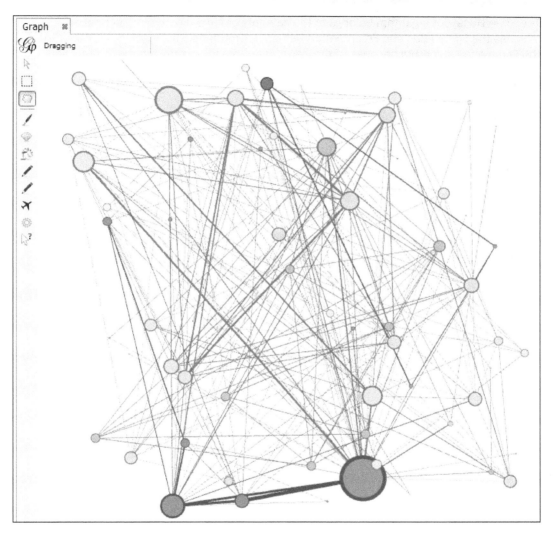

Using the Yifan Hu layout algorithm

The Yifan Hu layout algorithm belongs to the category of force-directed algorithms, which includes the Force Atlas and Fruchterman Reingold algorithms. This algorithm is faster than the Force Atlas algorithm because of the way it optimizes the overall internode repulsions in the network. The details of this algorithm will be discussed in the *How it works...* section of this recipe.

How to do it...

We will begin with the Les Misérables network and explain how to use the Yifan Hu layout algorithm to obtain a restructured network. So, let's get started:

1. Load the Les Misérables graph in Gephi.

2. In the **Layout** panel, click on the drop-down menu that says **---Choose a layout**.

3. From the drop-down menu, select **Yifan Hu**.

4. Hit **Run**. The restructured graph will appear in the **Graph** panel.

The following screenshot shows the Les Misérables graph after it has been restructured by using the Yifan Hu layout algorithm:

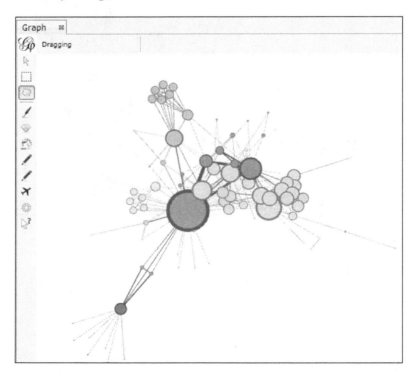

How it works...

The Yifan Hu layout algorithm uses the same concept as Force Atlas to compute the new layout of the network by optimizing the overall internode repulsions. The difference lies in the pair of nodes that are taken into consideration for the computation of repulsive forces. In the Yifan Hu layout algorithm, only pairs of adjacent nodes are taken into consideration. This is different from the Force Atlas algorithm in which every pair of nodes is considered for the computation of forces. This leads to reduced complexity in the Yifan Hu layout algorithm and hence the new layout is computed much faster.

The Yifan Hu layout algorithm also makes use of the concept of "adaptive cooling scheme". The algorithm starts with a high step length and over time readjusts the step length to smaller values. This has two positive implications. It helps in faster convergence of the algorithm and also in preventing the algorithm from getting stuck in local minima.

There's more...

There are several settings that the user can change for the Yifan Hu layout algorithm to come up with the best possible layout for the graph under consideration. Here are short descriptions of these settings:

- **Quadtree Max Level**: This defines the maximum level for the quadtree representation. A quadtree is a tree where each non-leaf node has exactly four children. The quadtree representation essentially places each node of the tree in a matrix, with four neighbors being its children. A higher value of this parameter results in higher accuracy and vice versa.

- **Theta**: This defines the approximation coefficient for the Barnes-Hut algorithm. A small value of theta would mean high accuracy.

- **Optimal Distance**: The edges in the graph in the Yifan Hu algorithm are visualized and assumed to be springs. This parameter defines the length of these springs. To get the nodes further apart from each other, use a large value for this parameter. To obtain a denser graph, use a small value.

- **Relative Strength**: This defines the ratio between the repulsive forces and the attraction forces in the graph.

- **Initial Step Size**: This is the initial step size that the algorithm will use in its integration phase. As prescribed in Gephi, a meaningful value, which is usually 10 percent of the optimal distance, should be chosen for this parameter.

- **Step ratio**: This is ratio that will be used to recompute and update the step size during the execution of the layout algorithm.

- **Adaptive cooling**: This option is for choosing the adaptive cooling scheme to configure the step size.

- **Convergence threshold**: This defines the threshold energy convergence levels for the algorithm to stop its execution. A high threshold will result in low accuracy, and a low threshold in high accuracy, in the resulting layout.

To reset all the values to their defaults, click on the **Reset** button located at the bottom of the left-hand side panel, adjacent to the **Presets** button.

See also

- *Efficient, High Quality Force-Directed Graph Drawing* by Yifan Hu, which was published in *The Mathematica Journal* in 2006, to learn more about the adaptive cooling scheme that has been introduced in this recipe. The paper can be accessed at http://www.mathematica-journal.com/issue/v10i1/contents/graph_draw/graph_draw.pdf.

- http://blog.notdot.net/2009/11/Damn-Cool-Algorithms-Spatial-indexing-with-Quadtrees-and-Hilbert-Curves for more information on quadtrees.

- An interactive d3.js implementation of quadtrees at http://bl.ocks.org/mbostock/4343214.

Using the Yifan Hu Proportional layout algorithm

This Yifan Hu Proportional layout algorithm is very much similar to the Yifan Hu layout algorithm, except that it uses a proportional displacement strategy for node placement in the graphical space. The accuracy and speed are almost comparable to that of Yifan Hu's.

How to do it...

We will begin with the Les Misérables network and explain how to use the Yifan Hu Proportional algorithm on it. The steps remain the same for any other network too. So let's get started:

1. Load the Les Misérables graph in Gephi.

2. In the **Layout** panel, click on the drop-down menu that says ---**Choose a layout**.

3. From the drop-down menu, select **Yifan Hu Proportional**.

4. Hit **Run**.

The following screenshot shows how the Les Misérables graph will look after the Yifan Hu Proportional algorithm has been executed:

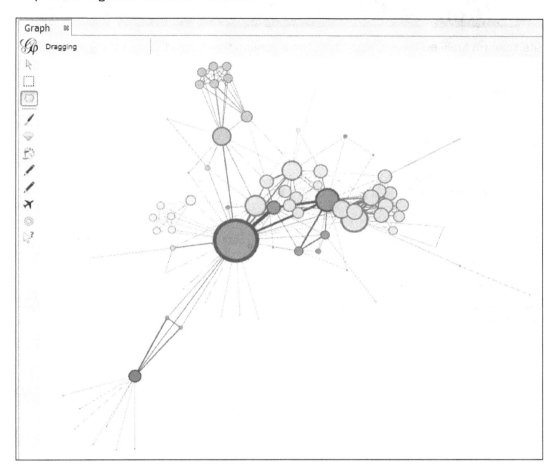

See also

▶ Yifan Hu's paper titled *Efficient, High-Quality Force-Directed Graph Drawing* at `http://www.mathematica-journal.com/issue/v10i1/contents/graph_draw/graph_draw.pdf` to read more about the Yifan Hu Proportional layout algorithm

Using the Yifan Hu Multilevel layout algorithm

The Yifan Hu Multilevel layout algorithm is an algorithm that brings together the good parts of force-directed algorithms and a multilevel algorithm to reduce algorithm complexity. This is one of the algorithms that works really well with large networks. In this recipe, we will see how this algorithm can be used to restructure the graphs.

How to do it...

In this recipe, we will learn how to use Yifan Hu Multilevel layout algorithm to obtain a restructured network on the Les Misérables network. The steps remain the same for any other network too. So, let's get started.

1. Load the Les Misérables graph in Gephi.

2. In the **Layout** panel, click on the drop-down menu that says ---**Choose a layout**.

3. From the drop-down menu, select **Yifan Hu Multilevel**.

4. Hit **Run**. The remodeled graph will appear in the **Graph** panel.

The following screenshot shows how the Les Misérables graph will look after the execution of the Yifan Hu Multilevel layout algorithm:

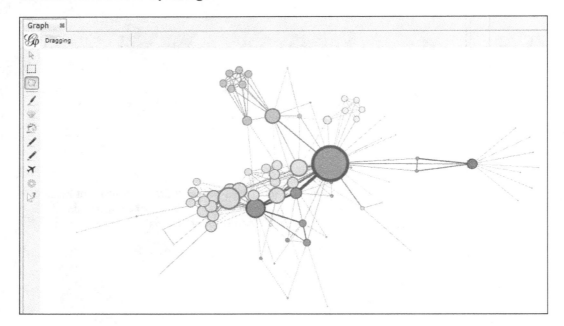

How it works...

The Yifan Hu Multilevel algorithm makes use of a scheme that combines a force-directed model and a graph-coarsening technique. Graph-coarsening tries to group vertices and build tighter, smaller graphs from these groups and is usually the first phase of a multilevel hierarchy building method. This results in initial attraction between nodes, followed by repulsion between them. It also results in reduced complexity of the algorithm. An approximation using the Barnes-Hut algorithm is done for computing the force that is exerted on a node from a group of distant nodes. The node treats all these distant nodes as a super node. One thing to notice here is that this algorithm doesn't use the edge weight for its computation.

There's more...

The settings for the Yifan Hu Multilevel algorithm can be modified, depending on the network in consideration. Here's a list of parameters that could be changed under **Settings** and a brief description for each of those parameters:

- ▶ **Quadtree max level**: This is similar to **Quadtree max level** described in *There's more...* section of the *Using the Yifan Hu layout algorithm* recipe.

- ▶ **Theta**: This is an approximation parameter to be used in the Barnes-Hut algorithm.

- ▶ **Minimum Level Size**: This defines the minimum number of nodes that each level of the quadtree must have.

- ▶ **Minimum Coarsening Rate**: This defines the minimum relative size between any two levels

- ▶ **Step Ratio**: This will be used to recompute and update the step size during execution of the layout algorithm.

- ▶ **Optimal Distance**: The edges in the graph in the Yifan Hu algorithm are visualized and assumed to be springs. This parameter defines the length of these springs. To get the nodes far apart from each other, use a large value for this parameter. To obtain a denser graph, use a small value

To reset all the values to their defaults, click on the **Reset** button, which is located at the bottom of the left-hand side panel, adjacent to the **Presets** button.

See also

- ▶ *Efficient, High-Quality Force-Directed Graph Drawing* by Yifan Hu at `http://www.mathematica-journal.com/issue/v10i1/contents/graph_draw/graph_draw.pdf` to learn more about the Yifan Hu Multilevel layout algorithm

4
Working with Partition and Ranking Algorithms

In this chapter, we will cover the following recipes:

- ▶ Partitioning the graph based on node attributes
- ▶ Partitioning the graph based on edge attributes
- ▶ Configuring node colors in a graph by ranking nodes
- ▶ Configuring node sizes in a graph by ranking nodes
- ▶ Configuring node label colors in a graph by ranking nodes
- ▶ Configuring node label sizes in a graph by ranking nodes
- ▶ Configuring edge colors in a graph by ranking edges
- ▶ Configuring the colors of edge labels in a graph by ranking edges
- ▶ Configuring the size of edge labels in a graph by ranking edges

Introduction

Clustering is an operation that is often performed while doing exploratory analysis on graphs or networks. Clustering refers to the process of classifying the nodes or edges of a graph in such a way that nodes or edges with similar attributes and properties get classified in the same bucket. Graph partitioning is synonymous with graph clustering.

Ranking nodes or edges in a graph is yet another operation that is widely performed during the exploratory analysis of graphs. Ranking could be done either on nodes or edges, depending on what the user is looking for.

This chapter will take you through the two processes of ranking and partitioning graphs based on user-defined metrics and modifying the graph visualization based on these parameters.

Partitioning the graph based on node attributes

In this recipe, we will learn how to partition the graph based on node attributes, which can be selected from the readily available Gephi GUI, and then assign the desired colors to the nodes to clearly visualize these partitions.

Getting ready

To get started, load the Les Misérables network in Gephi.

How to do it...

The following steps describe the process of partitioning the Les Misérables network, based on modularity class and then assigning colors to the nodes of each of the partitions, accordingly. The process remains the same for any other graph as well.

1. With the Les Misérables graph open in Gephi, click on the **Nodes** tab under the **Partition** panel, which is located on the upper-left side of the Gephi window.

2. Click on the refresh button with two encircling green arrows, which is located just below the **Nodes** tab marker.

3. This populates the drop-down menu with a list of available partitioning parameters:

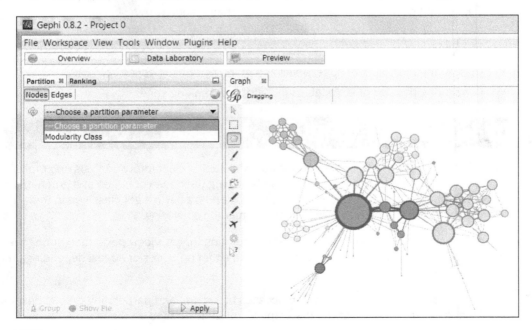

4. From the drop-down menu, select **Modularity Class**. A list of modularity classes for the Les Misérables graph will appear on the panel with some random colors allocated to them:

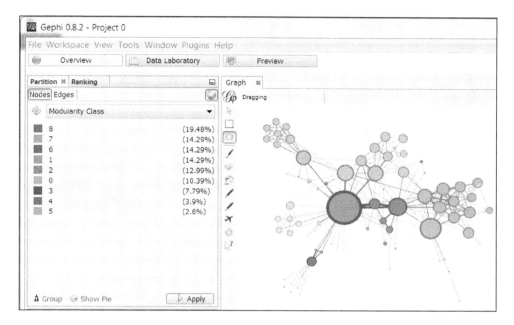

5. Click on **Apply** to assign these colors to the nodes in their respective modularity classes:

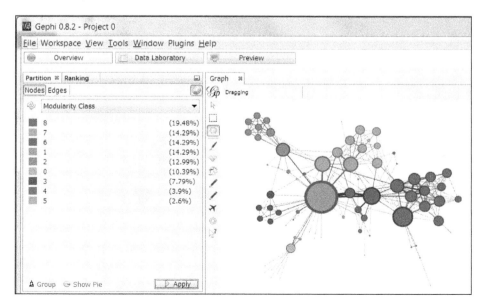

6. Click on **Group** in the lower-left side of the **Partition** panel to group the nodes that belong to the same modularity class into a single node:

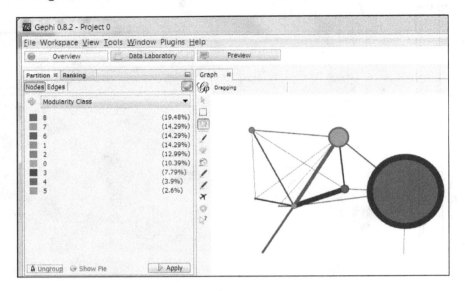

7. You can also change the colors separately for each modularity class. To do so, click and hold the color box next to the modularity class and drag the mouse pointer to the desired shade. Click on **Apply**.

The following screenshot shows how the Les Misérables network will look if black was assigned to modularity classes 0, 1, 2, 3, and 4, and red was assigned to the remaining classes:

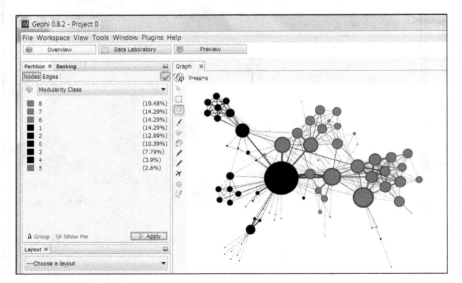

8. You can choose to switch between the different color palettes available in Gephi to pick the color of your choice. To do so, place the mouse pointer over the color box next to modularity class, hold the *Ctrl + Shift + Alt* keys, then click and hold to make a different color palette appear.

How it works...

Modularity is used to identify clusters or communities, as clusters are popularly called in the networks world, in a given graph. Modularity results in grouping of nodes that are far more strongly connected than they would have been in a random graph. Modularity gives an insight into the strength of a given network. A high modularity score for the graph would mean that different parts or modules of the graph are strongly connected to each other and, hence, the graph is more robust.

Modularity is useful when studying a network's robustness. One such application is finding out how robust a communication network of hubs and switches is against hacker attacks.

There's more...

Modularity isn't the only partitioning parameter present in Gephi. There are many more parameters in Gephi, which are listed under the **Settings** panel, that could be used to partition the graph. One such parameter is degree. The degree of a node of a graph refers to the number of nodes this node is connected to.

In the case of directed graphs, the degree could either be in-degree or out-degree. In-degree refers to the number of inbound connections to the node and out-degree refers to the number of outbound connections from it. Of course, the logical partitions that are created by partitioning the graph according to the modularity score are different from those created according to the degree. Modularity score groups nodes by their connections, thereby creating clusters of related nodes. On the other hand, a degree will not necessarily create clusters of related nodes but only clusters of nodes with similar quantitative attributes. These nodes may not be related to each other otherwise.

The following steps take you through the process of partitioning the graph based on node degree:

1. Load the Les Misérables network in Gephi.

2. In the **Settings** panel, located on the right side of the Gephi window, click on **Run** against **Average Degree**:

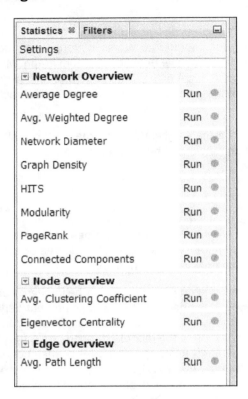

3. This opens up a pop-up window and runs the average degree metrics on the Les Misérables network:

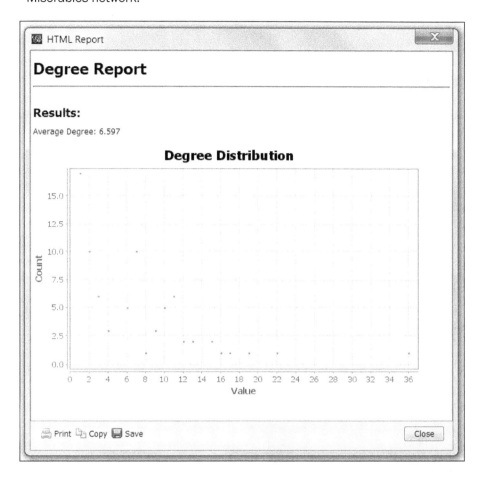

4. Hit **Close** to close this pop-up window. You can choose to save or print out these results by using the respective options available in the pop-up window.

5. Now, under the **Nodes** tab of the **Partition** panel, click on the Refresh button to repopulate the list of available partitioning parameters:

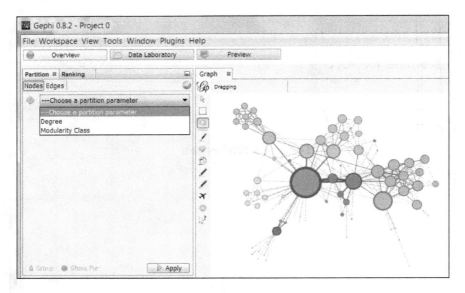

6. Choose **Degree** from the drop-down list. This populates the **Partition** panel with a list of multiple degrees present in the graph and allocates random colors to each of them.

7. Hit **Apply** to assign these colors to the respective nodes in the graph. The following screenshot shows the Les Misérables network when partitioned on its node degrees:

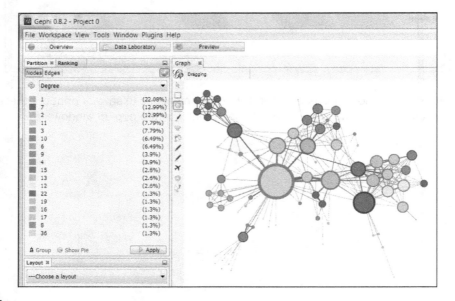

See also

▶ http://en.wikipedia.org/wiki/Modularity_(networks) for more information on modularity.

▶ *Fast unfolding of communities in large networks* by Vincent D. Blondel, Jean-Loup Guillaume, Renaud Lambiotte, and Etienne Lefebvre, which was published in the October 2008 edition of *Journal of Statistical Mechanics: Theory and Experiment*, at http://arxiv.org/pdf/0803.0476.pdf to get a deeper understanding into the concept of modularity.

▶ *Information Epidemics and Synchronized Viral Social Contagion* by Dmitry Paranyushkin to understand about the application of modularity in social networks. The paper and some related videos can be found at http://noduslabs.com/research/information-epidemics-viral-social-contagion/.

Partitioning the graph based on edge attributes

Nodes are the entities in a graph and edges depict the relationships among these nodes. Sometimes, more than the nodes, these edges are of more interest to anyone performing exploratory analysis on graphs. In this recipe, we will learn how to partition the graph based on edge parameters and color the graph accordingly, in order to get better insights into the network under consideration.

Getting ready

In order to get started with this recipe, we will need to load the Les Misérables network into Gephi. Gephi doesn't have an off-the-shelf partitioning technique that utilizes an edge-based parameter. So, go to the **Data Laboratory** mode in Gephi, click on the **Edges** tab in the **Data Table** panel, and add labels for some or all of the edges. Return to **Overview** mode.

How to do it...

The following steps describe how to partition the graph, based on the label classes of the edges of the Les Misérables network. These are the labels that we have just added:

1. Click on the **Edges** tab in the **Partition** panel.

2. Click on the refresh button to see the list of edge parameters available. Choose **Label** from the drop-down list:

3. This populates the panel with a table of label classes and specific colors assigned to these classes randomly. To assign a different color to a class, place the mouse pointer on the color box for that modularity class, click and hold the mouse button in place, and then drag the pointer to the desired shade.

The following screenshot shows the Les Misérables network with its edges recolored according to the respective edge labels:

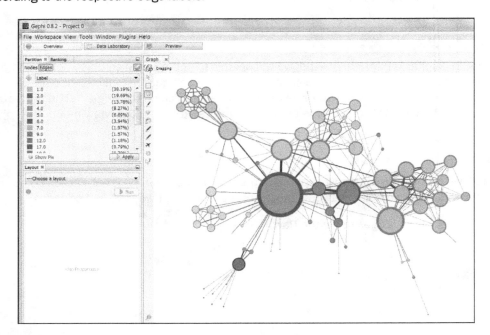

There's a possibility that the edges may not be recolored correctly. To solve this problem, click on the upward pointing arrow located in the lower-right corner of the **Graph** panel to expand the bottom toolbar. Click on the **Edges** tab and uncheck **Source Node Color**. Hit **Apply** once again in the **Partition** panel to recolor correctly this time.

How it works...

Edge labels play the role of differentiators between various parts of a graph. They usually contain information that relates one edge to another or distinguishes one edge from another. For example, in the case of a professional social network such as LinkedIn, edges might contain information about the type of connection between two people. The various connections can be ex-classmate connections, past-colleague connections, current-colleague connections, and so on. Once these labels are assigned to the edges, one can partition the graph according to these labels by assuming each label to be a separate partition.

There's more...

You can also define multiple edge attributes for edges in a given graph and create multiple edge parameters for partitioning. This can be done by adding an extra row in the **Edges** datasheet in the **Data Laboratory** mode. This row will define a new edge attribute for this graph with user-entered values. For example, the weight of the edges is a very important attribute when it comes to real-world networks and complex networks. For a basic graph, the default edge weight is 1 but this will not be the case for real-world networks. The edge weight, for example, helps in differentiating between strong and weak ties. Hence, it is very often useful to have the edge weight as one of the attributes. The details of this procedure will be described later in *Chapter 6, Working in the Data Laboratory Mode*.

Configuring node colors in a graph by ranking nodes

Gephi allows users to reset node properties such as node color and node size in a graph, according to a continuous attribute. This process is termed "ranking" and can happen only for continuous attributes. For example, if we were to consider a modified version of the Les Misérables graph where information such as age and gender is present for every character, then ranking can be performed for the age attribute, whereas, when carrying out graph partition, we will use an attribute such as gender. In this recipe, we will learn how to reassign colors to nodes of a graph, depending on a continuous node attribute.

Getting ready

To get started with this recipe, load the Les Misérables graph in Gephi.

How to do it...

Let's recolor the nodes of the Les Misérables graph by using the following steps:

1. Click on the **Nodes** tab in the **Ranking** panel that is located next to the **Partition** panel on the upper-left side of the Gephi application window.

2. Click on the multi-colored circular icon in the upper-right corner of the **Ranking** panel as shown in the following screenshot.

3. From the drop-down, choose a rank parameter such as **Modularity Class**:

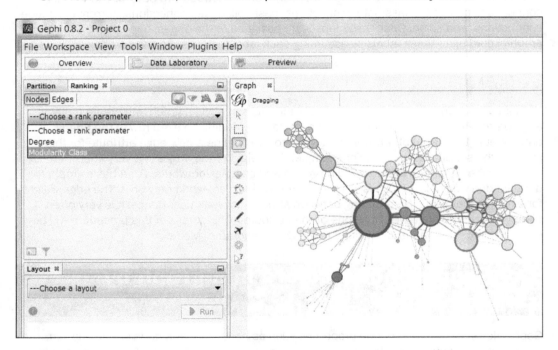

4. Click on the small, square-shaped box located on the upper-right corner of the table that has just appeared and select the desired color palette:

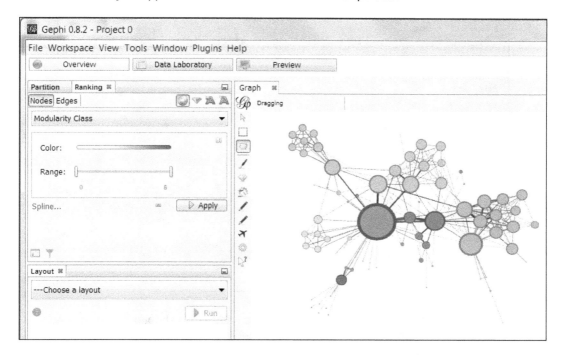

5. Move the cursor over the **Color** slider to change the color gradient.

6. Move the **Range** slider to configure which nodes have to be recolored and which nodes have to be reset to white.

7. Click on **Spline** to select the distribution of the color gradient in this process.

8. Hit **Apply**. The following screenshot shows the Les Misérables network ranked according to modularity class:

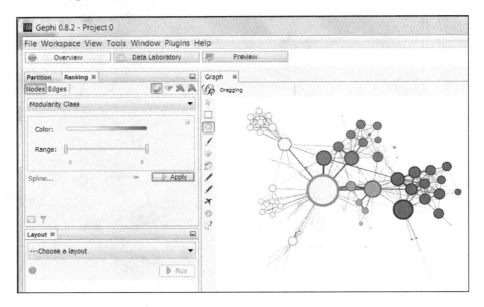

9. To see the results of this process, click on the small table-shaped button located in the lower left-most button of the **Ranking** panel. This will open a table with a list of modularity classes along with their ranks and colors:

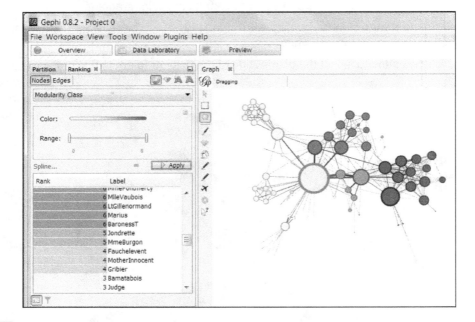

How it works...

Gephi's ranking algorithm assigns a rank to each of the nodes in the graph based on a continuous attribute such as degree. It then assigns a continuous palette of colors to these ranked nodes.

See also

▶ *Ranking on Graph Data* by Shivani Agarwal that appeared in the *Proceedings of International Conference on Machine Learning* in 2006 for a more mathematical treatment of the ranking of graphs. The paper is accessible at `http://dl.acm.org/citation.cfm?id=1143848`.

Configuring node sizes in a graph by ranking nodes

In the previous recipe, we learned how to configure node colors by ranking the graph based on a continuous attribute. In this recipe, we will learn how to configure node sizes by ranking the graph under consideration based on a continuous attribute.

How to do it...

The following steps illustrate the process of altering node sizes by ranking a graph based on a continuous node attribute:

1. Click on the **Nodes** tab in the **Ranking** panel.
2. Click on the red diamond-shaped icon in the upper-right corner of the **Ranking** panel.

3. From the drop-down, choose a rank parameter such as **Degree**:

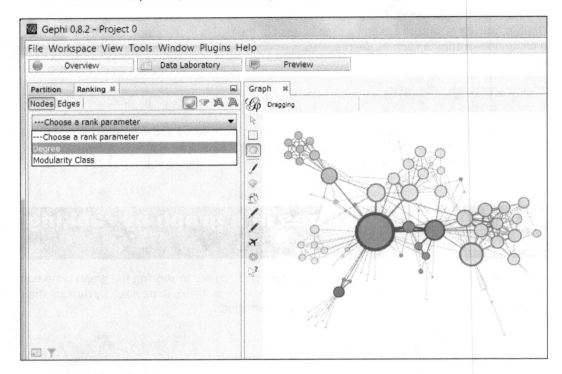

4. Click on the diamond-shaped icon located in the upper-right corner of the **Ranking** panel.

5. Enter the minimum and maximum allowed node size in the **Min Size** and **Max Size** input boxes, respectively.

6. Move the **Range** slider to configure which nodes have to be resized and which nodes are to retain their original size.

7. Click on **Spline** to select the distribution of the size gradient in this process.

8. Hit **Apply**. The following screenshot shows the Les Misérables network ranked according to degree class:

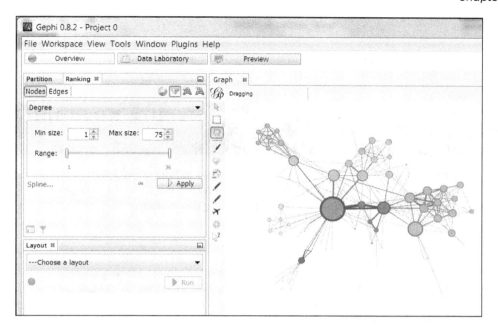

9. To see the results of this process, click on the small table-shaped button located in the lower left-most corner of the **Ranking** panel. This will open a table with a list of modularity classes, along with their ranks:

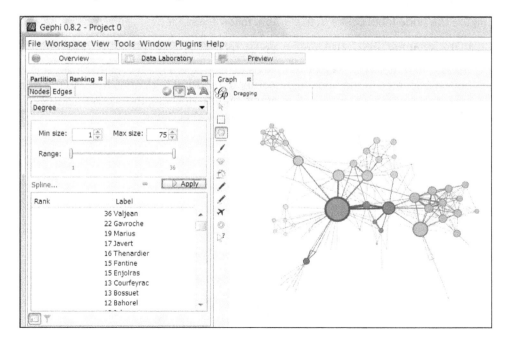

How it works...

This recipe works pretty much in the same way as the one that involves altering the node colors by ranking the graph.

Configuring node label colors in a graph by ranking nodes

Node label colors can also be configured in a similar way to how node colors and sizes can be configured by ranking the nodes of the graph. This recipe will describe this process of configuring node label colors, based on various ranking parameters.

Getting ready

Load the Les Misérables graph in Gephi. Click on the small upward-facing triangle in the bottom-right corner of the **Graph** panel to expand the bottom panel. In the **Labels** tab of this panel, check the checkbox next to **Node** to make the node labels appear on the graph. To make sure that the final visualization is neat and comprehensible, run the Fruchterman Reingold algorithm, followed by the Label Adjust algorithm on this graph.

How to do it...

To configure the node label colors in this graph, follow these steps:

1. In the **Ranking** panel, click on the **Nodes** tab.

2. Select the button that has a capital A as its icon and a multi-colored circle. This button is located in the upper-right corner of the **Ranking** panel. Refer to the following screenshot to identify the location of this button. Hovering the mouse pointer on top of this button should open a popup that reads **Label Color**:

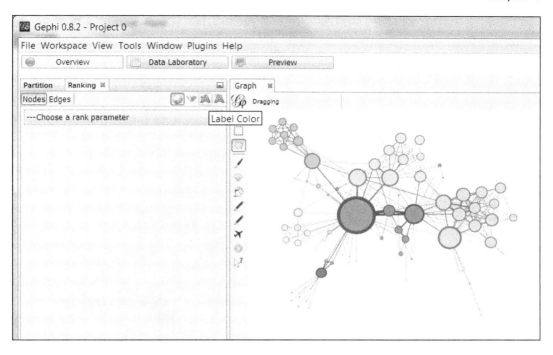

3. Choose a ranking parameter from the drop-down list.

4. Fix the color palette and color gradient for the range of colors to be assigned to these labels.

5. Move the slider to descope the nodes belonging to a specific range of the selected ranking parameter. The colors of these node labels will not be affected as a result of this operation.

6. Click on the small table-shaped icon that is located at the lower-left corner of the **Ranking** panel to view details of the execution of this operation.

7. Hit **Apply**.

The following screenshot shows the Les Misérables graph when visualized after running the Fruchterman Reingold and Label Adjust algorithms on it, followed by recoloring the labels according to the ranking parameter as a degree of its nodes:

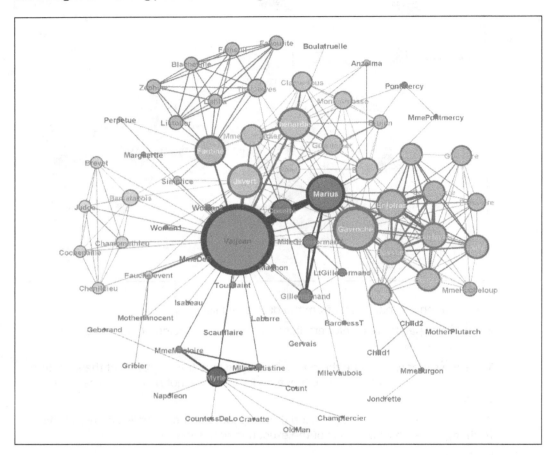

How it works...

This recipe works pretty much in the same way as the one that involves altering the node colors by ranking the graph. The only difference is that here we are coloring the node labels instead of the nodes themselves.

Configuring node label sizes in a graph by ranking nodes

One can choose to configure node label sizes rather than their colors. In this recipe, we explain how to achieve this by ranking nodes.

Getting ready

Load the Les Misérables graph in Gephi. Click on the small upward-facing triangle in the bottom-right corner of the **Graph** panel to expand the bottom panel. In the **Labels** tab in this panel, check the checkbox next to **Node** to make the node labels appear on the graph. To make sure that the final visualization is neat and comprehensible, run the Fruchterman Reingold algorithm, followed by the Label Adjust algorithm, on this graph.

How to do it...

To adjust the sizes of node labels by using the node ranking for a graph, follow these steps:

1. In the **Ranking** panel, click on the **Nodes** tab.

2. Select the button that has its icon as capital letter **A** and a red diamond situated on top of the letter A. This button is located in the upper-right corner of the **Ranking** panel. Refer to the following screenshot to identify the location of this button. Hovering the mouse pointer on top of this button should open a popup that reads **Label Size**:

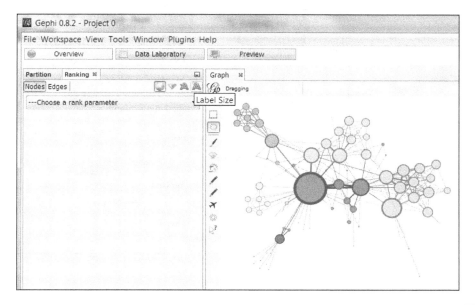

3. Choose a ranking parameter from the drop-down list.

4. Fix the minimum and maximum size for the node labels.

5. Move the slider to descope the nodes belonging to a specific range of the selected ranking parameter. The colors of the node labels will not get affected as a result of this operation.

6. Click on the small table-shaped icon located in the lower-left corner of the Ranking panel to view details of the execution of this operation.

7. Hit **Apply**.

The following screenshot shows the Les Misérables graph, when visualized after running the Fruchterman Reingold and Label Adjust algorithms on it followed by resizing the labels according to the ranking parameter as a degree of its nodes:

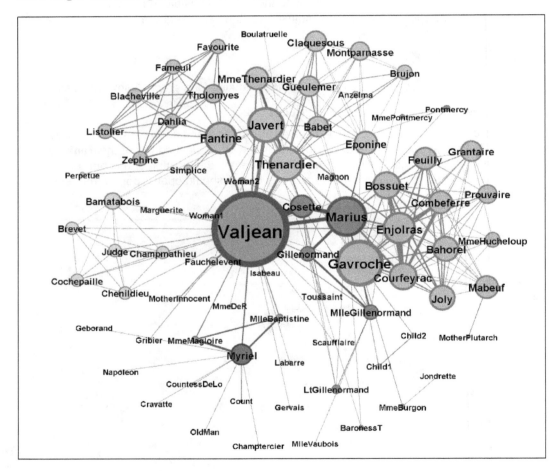

How it works...

This recipe works in pretty much the same way as the one that involves altering the node colors by ranking the graph. The only difference is that we are coloring the node labels here, instead of the nodes themselves.

Configuring edge colors in a graph by ranking edges

So far, we have only seen recipes that configure the properties of the nodes, such as color and size, in a graph. We will now move on to learning how to configure edge properties, such as edge color, depending on the ranking of the graph.

How to do it...

The following steps describe how to configure edge colors in a graph by ranking the edges in the graph:

1. Load the Les Misérables network into Gephi.

2. In the **Ranking** panel, click on the **Edges** tab.

3. Click on the button with a multi-colored circle as its icon. This button is located in the upper-right corner of the **Ranking** panel adjacent to the **Edges** tab. Hovering the mouse pointer on top of this button should result in a popup that reads **Color**.

4. Choose a ranking parameter from the drop-down list.

5. Choose the desired color palette and fix color gradient in the same way as you did in all the previous recipes described in this chapter.

6. Move the slider to descope the nodes belonging to a specific range of the selected ranking parameter. The colors of the node labels will not get affected as a result of this operation.

7. Click on **Spline** to configure rank interpolation and choose from the given options.

8. Click on the small table-shaped icon located in the lower-left corner of the **Ranking** panel to view details of the execution of this operation.

9. Hit **Apply**.

The following screenshot shows the Les Misérables graph with its edges recolored after ranking with edge weight as the ranking parameter:

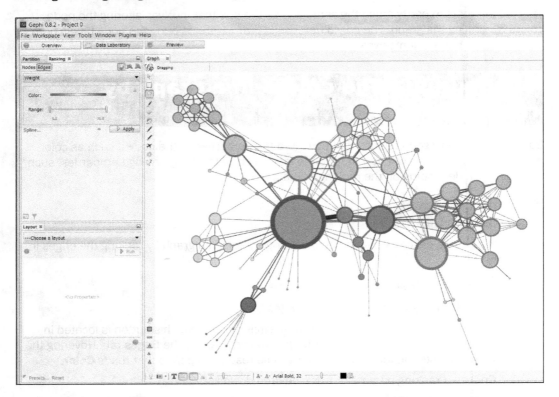

How it works...

The algorithm assigns one shade from the chosen color palette to one degree class. This happens recursively until all the classes resulting from the ranking operation on the graph have been assigned colors.

Configuring the colors of edge labels in a graph by ranking edges

Earlier in this chapter, we learned how to reconfigure the colors of node labels in a graph by ranking the graph. In this recipe, we will learn how to perform the same operation for edge labels.

Getting ready

Load the Les Misérables graph in Gephi. Click on the small upward-facing triangle in the bottom-right corner of the **Graph** panel to expand the bottom panel. In the **Labels** tab of this panel, check the checkbox next to **Edge**. Now click on **Configure** and then click on **Edge**. Select the checkbox next to **Weight** and hit **OK** to make the node labels appear on the graph. To make sure that the final visualization is neat and comprehensible, run the Fruchterman Reingold algorithm, followed by the Label Adjust algorithm, on this graph.

How to do it...

To configure the colors of edge labels by ranking edges in a graph, follow these steps:

1. Load the Les Misérables network into Gephi.
2. In the **Ranking** panel, click on the **Edges** tab.
3. Select the button that has a capital letter **A** as its icon and a multi-colored circle. This button is located in the upper-right corner of the **Ranking** panel. Hovering the mouse pointer on top of this button should open a popup that reads **Label Color**.
4. Choose a ranking parameter from the drop-down list.
5. Fix the color palette and color gradient for the range of colors to be assigned to these labels.
6. Move the slider to descope the nodes belonging to a specific range of the selected ranking parameter. The colors of these node labels will not get affected as a result of this operation.
7. Click on the small table-shaped icon, which is located in the lower-left corner of the **Ranking** panel, to view details of the execution of this operation.
8. Hit **Apply**.

The following screenshot shows the Les Misérables graph, when visualized after running the Fruchterman Reingold and Label Adjust algorithms on it followed by recoloring the edge labels according to the ranking parameter as a weight of its edges:

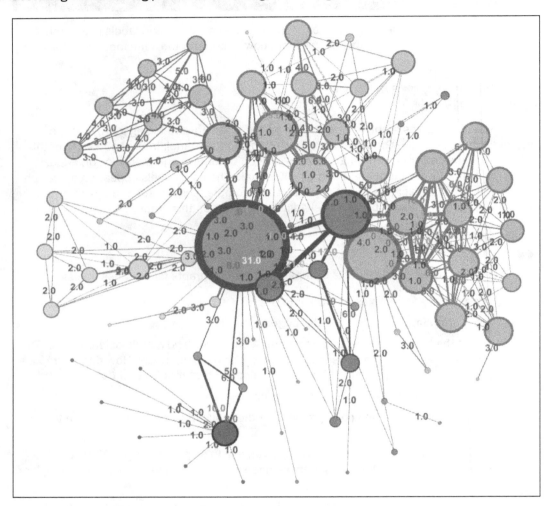

How it works...

This recipe works in a similar way to reconfiguring the colors of node labels by ranking nodes in the graph. The only difference is that, in this case, the ranking takes place on the edges instead of nodes.

Configuring the size of edge labels in a graph by ranking edges

The previous recipe described how to reconfigure the colors of edge labels by ranking edges in a graph. It is also possible to reconfigure the size of the edge labels by ranking edges in a graph, and that is exactly what this recipes teaches.

Getting ready

Load the Les Misérables graph in Gephi. Click on the small upward-facing triangle placed on the bottom-right corner of the **Graph** panel to expand the bottom panel. In the **Labels** tab of this panel, check the checkbox next to **Node** to make the node labels appear on the graph. To make sure that the final visualization is neat and comprehensible, run the Fruchterman Reingold algorithm, followed by the Label Adjust algorithm, on this graph.

How to do it...

To configure the sizes of edge labels by ranking edges in a graph, follow these steps:

1. In the **Ranking** panel, click on the **Edges** tab.
2. Select the button that has a capital letter **A** as its icon and a red diamond on top of the letter A. This button is located in the upper-right corner of the **Ranking** panel. Hovering the mouse pointer on top of this button should open a popup that reads **Label Size**.
3. Choose an edge ranking parameter from the drop-down list.
4. Fix the color palette and color gradient for the range of colors to be assigned to these labels.
5. Move the slider to de-scope the nodes belonging to a specific range of the selected ranking parameter. The colors of these nodes' labels will not be affected as a result of this operation.
6. Click on the small table-shaped icon located at the lower-left corner of the **Ranking** panel to view details of the execution of this operation.
7. Hit **Apply**.

The following screenshot shows the Les Misérables graph when visualized after running the Fruchterman Reingold and Label Adjust algorithms on it, followed by resizing the edge labels according to the ranking parameter as a weight of its edges:

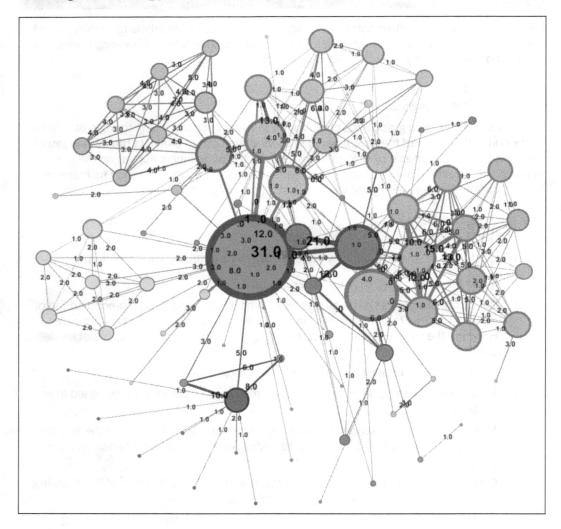

5
Running Metrics, Filters, and Timelines

In this chapter, we will cover the following recipes:

- ▸ Selecting a list of metrics for a graph
- ▸ Finding the average degree and average weighted degree of a graph
- ▸ Finding the network diameter
- ▸ Finding graph density
- ▸ Finding the HITS value for a graph
- ▸ Finding a graph's modularity
- ▸ Finding a graph's PageRank
- ▸ Finding connected components in a graph
- ▸ Getting a node overview of a graph
- ▸ Getting an edge overview of a graph
- ▸ Getting dynamic statistics for a graph
- ▸ Applying individual filters on a graph
- ▸ Applying a combination of filters on a graph
- ▸ Filtering dynamic graphs based on time intervals

Introduction

Gephi provides some ready-to-use ways to study the statistical properties of graphical networks. These statistical properties include the properties of the network as a whole, as well as individual properties of nodes and edges within the network. This chapter will enable readers to learn some of these properties and how to use Gephi to explore them. So let's get started!

Selecting a list of metrics for a graph

Gephi offers a wide variety of metrics for exploring graphs. These metrics allow users to explore graphs from various perspectives. In this recipe, we will learn how to access these different metrics for a specified graph.

Getting ready

Load a graph of your choice in Gephi.

How to do it...

To view different metrics available in Gephi for exploring a graph, follow these steps:

1. In the **Statistics** panel situated on the right-hand side of the Gephi window, find the tab that reads **Settings**.

2. Click on the **Settings** tab to open up a pop-up window.

3. From the list of available metrics in the pop-up window, check the ones that you would like to work with:

4. Click on **OK**. The **Statistics** panel will get populated with the selected metrics, as shown in the following screenshot:

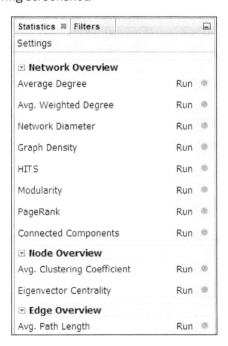

Finding the average degree and average weighted degree of a graph

The degree of a node in a graph is defined as the number of edges that are incident on that node. The loops—that is, the edges that have the same node as their starting and end point—are counted twice. In this recipe, we will learn how to find the average degree and average weighted degree for a graph.

How to do it...

The following steps illustrate the process to find the average degree and weighted degree of a graph:

1. Load or create a graph in Gephi. For this recipe, we will consider the Les Misérables network that's already available in Gephi and can be loaded at the **Welcome** screen.

2. In the **Statistics** panel located on the right-hand side of the Gephi application window, under the **Network Overview** tab, click on the **Run** button located beside **Average Degree**:

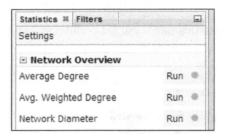

3. This opens up a window containing the degree report for the Les Misérables network, as shown in the following screenshot. In the case of directed graphs, the report contains the in-degree and out-degree distributions as well:

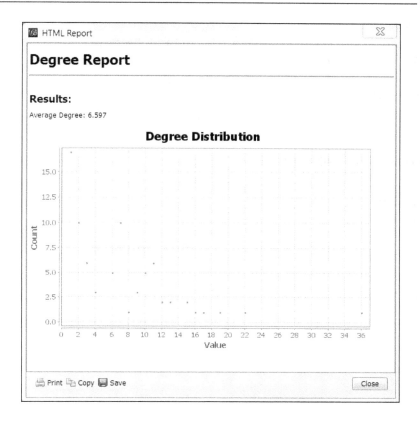

The graph in the preceding screenshot depicts the degree distribution for the Les Misérables network. This pop-up window has options for printing, copying, and/or saving the degree report.

The average degree of the Les Misérables network is now displayed in the **Statistics** panel beside the **Run** button for **Average Degree**, as shown in the following screenshot:

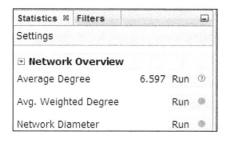

4. To find the average weighted degree of the Les Misérables graph, hit the **Run** button adjacent to **Avg. Weighted Degree** in the **Network Overview** tab of the **Statistics** panel in the Gephi window.

   This will open up a window containing the weighted degree report of the Les Misérables network, as shown in the following screenshot:

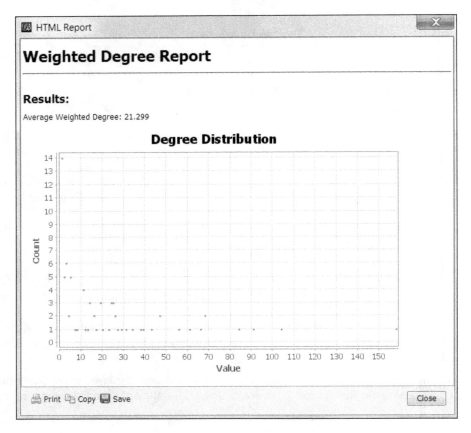

The average weighted degree of the Les Misérables graph is now also displayed in the **Statistics** panel that is adjacent to the **Run** button for **Avg. Weighted Degree**:

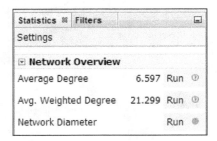

How it works...

The average degree for a graph is the measure of how many edges there are in the graph compared to its number of vertices. To find out the average degree for a graph, Gephi computes the sum of the degrees of individual nodes in the graph and divides that by the number of nodes present in it.

To find the average weighted degree for a graph with weighted edges, Gephi computes the average mean of the sum of the weights of the incident edges on all the nodes in the graph.

There's more...

If you have closed the report window and wish to see it once again, click on the small button with a question mark adjacent to the **Run** button. This will reopen the degree report.

See also

► The paper titled *Statistical Analysis of Weighted Networks* by Antoniou Ioannis and Tsompa Eleni (http://arxiv.org/ftp/arxiv/papers/0704/0704.0686.pdf) for more information about the statistical properties, such as average degree and weighted average of weighted networks

► An example of the applications of average degree and weighted average degree described by Gautam A. Thakur on his blog titled *Average Degree and Weighted Average Degree Distribution of Locations in Global Cities* at `http://wisonets.wordpress.com/2011/12/16/average-degree-and-weighted-average-degree-distribution-of-locations-in-global-cities/`

► Another explanation on the topic present in Matthieu Totet's blog at `http://matthieu-totet.fr/Koumin/2013/12/16/understand-degree-weighted-degree-betweeness-centrality/`

Finding the network diameter

The diameter of a network refers to the length of the longest of all the computed shortest paths between all pair of nodes in the network.

How to do it...

The following steps describe how to find the diameter of a network using the capabilities offered by Gephi:

1. Click on **Window** in the menu bar located at the top of the Gephi window. From the drop-down, select **Welcome**. Click on **Les Miserables.gexf**.

2. In the pop-up window, select **Graph Type** as **Directed**. This opens up the directed version of the Les Misérables network into Gephi.

3. In the **Statistics** panel, under the **Network Overview** tab, click on the **Run** button, which is next to **Network Diameter**, to open the **Graph Distance settings** window:

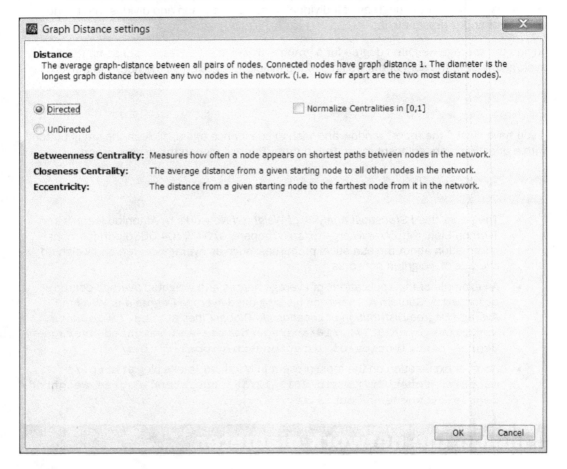

4. In the **Graph Distance settings** window, you can decide on which type of graph, **Directed** or **UnDirected**, the diameter algorithm has to be run. If you have loaded an undirected graph, the **Directed** radio button will remain deactivated. If a directed graph is chosen, you can choose between the directed and undirected versions of it to find the diameter.

5. Check the box next to **Normalize Centralities in [0, 1]** to allow Gephi to normalize the three centralities' values between zero and one. The three centralities being referred to here are **Betweenness Centrality**, **Closeness Centrality**, and **Eccentricity**.

6. Click on **OK**. This opens up the **Graph Distance Report** window, as displayed in the following screenshot, that shows the value of the network diameter, network radius, average path length, number of shortest paths, and three separate graphs depicting betweenness centrality distribution, closeness centrality distribution, and eccentricity distribution:

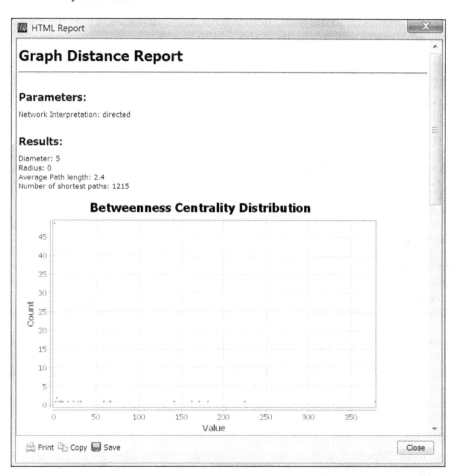

How it works...

The diameter of a network gives us the maximum number of hops that must be made to travel from one node in the graph to the other. To find the diameter, all the shortest paths between every pair of nodes in the graph are computed and then the length of the longest of them gives us the diameter of the network. If the network is disconnected—that is, if the network has multiple components that do not share an edge between them—then the diameter of such a network is infinite.

Note that, in the case of weighted graphs, the longest path that determines the diameter of the graph is not the actual length of the path but the number of hops that would be required to traverse from the starting vertex to the end vertex.

The computation of the diameter of a graphical network makes use of a property called the eccentricity of nodes. The eccentricity of a node is a measure of the number of hops required to reach the farthest node in the graph from this node. The diameter is then the maximum eccentricity among all the nodes in the graph.

There's more...

There are three concepts—betweenness centrality, closeness centrality, and eccentricity—that have been introduced in this recipe. Eccentricity has already been covered in the *How it works...* section of this recipe. Betweenness centrality and closeness centrality are yet more important statistical properties of a network and are applied in a lot of real-world problems such as finding influential people in a social network, finding crucial hubs in a computer network, finding congestion nodes in wireless networks, and so on.

The betweenness centrality of a node is an indicator of its centrality or importance in the network. It is described as the number of shortest paths from all the vertices to all the other vertices in the network that pass through the node in consideration.

The closeness centrality of a node measures how accessible every other node in the graph is from the considered node. It is defined as the inverse of the sum of shortest distances of every other node in the network from the current node. Closeness centrality is an indicator of the speed at which information will transfuse into the network, starting from the current node.

Yet another concept that has been mentioned in this recipe is the radius of the graph. The radius of a graph is the opposite of its diameter. It is defined as the minimum eccentricity among the vertices of the graph. In other words, it refers to the minimum number of hops that are required to reach from one node of the graph to its farthest node.

See also

▸ *A Faster Algorithm for Betweenness Centrality** by Ulrik Brandes to know more about the algorithm that Gephi uses to find the betweenness centrality indices. It was published in *Journal of Mathematical Sociology* in 2001 and can be found at http://www.inf.uni-konstanz.de/algo/publications/b-fabc-01.pdf.

▸ *Distance in Graphs* by Wayne Goddard and Ortrud R. Oellermann at http://citeseerx.ist.psu.edu/viewdoc/summary?doi=10.1.1.221.6262 for detailed information on the paths in a graph.

▸ *Social Network Analysis, A Brief Introduction* at http://www.orgnet.com/sna.html for information on various centrality measures in social networks

▶ *The Betweenness Centrality Of Biological Networks* at `http://scholar.lib.vt.edu/theses/available/etd-10162005-200707/unrestricted/thesis.pdf` to understand about the applications of betweenness centrality in biological networks.

▶ The book titled *Introduction to Graph Theory* by Douglas B. West to understand path lengths and centralities in graphs in detail.

Finding graph density

One another important statistical metric for graphs is density. In this recipe, you will learn what graph density is and how to compute it in Gephi.

How to do it...

The following steps illustrate how to use Gephi to figure out the graph density for a chosen graph:

1. Load the directed version of the Les Misérables network in Gephi, as described in the *How to do it...* section of the previous recipe.

2. In the **Statistics** panel located on the right-hand side of the Gephi application window, click on the **Run** button that is placed against **Graph Density**.

3. This opens up the **Density settings** window, as shown in the following screenshot, where you can choose between the directed or the undirected version of the graph to be considered for the computation of graph density:

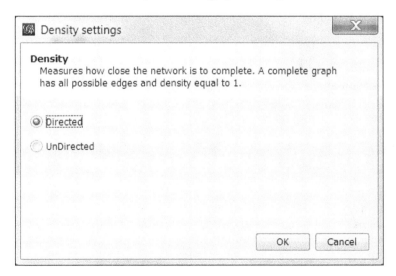

4. Click on **OK**. This opens up the following **Graph Density Report** window:

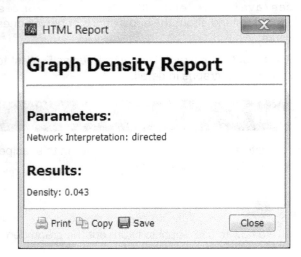

How it works...

A complete graph is a graph in which every pair of nodes is connected via a direct edge. The density of a graph is a measure of how close the graph is to a complete graph with the same number of nodes. It is defined as the ratio of the total number of edges present in a graph to the total number of edges possible in the graph. The total number of edges possible in a simple undirected graph is mathematically computed as $(N(N-1))/2$, where N is the number of nodes in the graph. A simple graph is a graph that has no loops and not more than one edge between the same pair of nodes.

There's more...

The density of the undirected version of a graph with n nodes will be twice of that of the directed version of the graph. This is because, in a directed graph, there are two edges possible between every pair of nodes, each with a different direction.

Finding the HITS value for a graph

Hyperlink-Induced Topic Search (**HITS**) is also known as **hubs** and **authorities**. It is a link analysis algorithm and is used to evaluate the relationship between the nodes in a graph. This algorithm aims to find two different scores for each node in the graph: authority, which indicates the value of the information that the node holds, and hub, which indicates the value of the link information to the other linked nodes to this node. In this recipe, you will learn about HITS and how Gephi is used to compute this metric for a graph.

How to do it...

Considering the directed version of the Les Misérables network, the following steps describe the process of determining the HITS score for a graph in Gephi:

1. In Gephi's menu bar, click on **Window**. From the drop-down menu, select **Welcome**.
2. In the window that just opened, click on **Les Miserables.gexf**. This opens up another window.
3. In the **Import Report** window, select **Graph Type** as **directed**.
4. With the directed version of Les Misérables loaded in Gephi, click on the **Run** button placed next to **HITS** in the **Network Overview** tab of the **Statistics** panel.
5. This opens up the **HITS settings** window, as shown in the following screenshot:

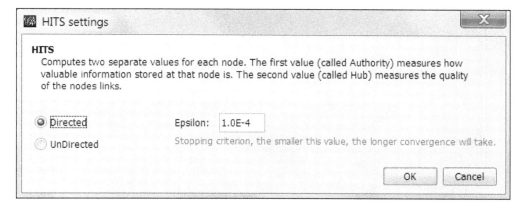

6. Choose the graph type, **Directed** or **UnDirected**, on which you would want to run the HITS algorithm.
7. Enter the stopping criterion value in the **Epsilon** textbox. This determines the stopping point for the algorithm.

8. Hit **OK**. This opens up **HITS Metric Report** with a graph depicting the hub and authority distribution for the graph:

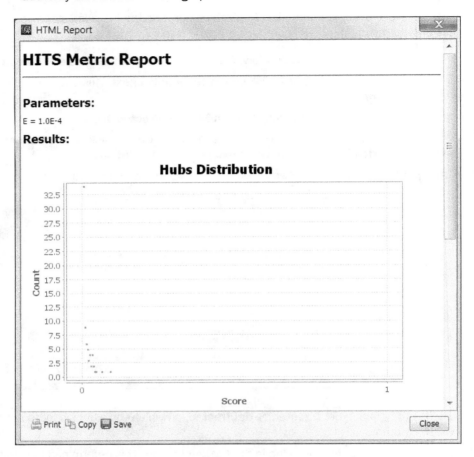

How it works...

The HITS algorithm was developed by Professor Jon Kleinberg from the department of computer science at Cornell University at around the same time as the PageRank algorithm was being developed. The HITS algorithm is a link analysis algorithm that helps in identifying the crucial nodes in a graph. It assigns two scores, a hub score and an authority score, to each of the nodes in the graph. The authority score of a node is a measure of the amount of valuable information that this node holds. The hub score of a node depicts how many highly informative nodes or authoritative nodes this node is pointing to. So a node with a high hub score shows that this node is pointing to many other authoritative nodes and hence serves as a directory to the authorities. On the other hand, a node with a high authoritative score shows that it is pointed to by a large number of nodes and hence serves as a node of useful information in the network.

One thing that you might have noticed is the Epsilon or stopping criterion for the HITS algorithm being mentioned in one of the steps of the recipe. Computation of HITS makes use of matrices and something called **Eigenvalues**. The value of Epsilon instructs the algorithm to stop when the difference between eigenvalues of the matrices for two consecutive iterations becomes negligibly small. The detailed discussion of eigenvalues and any mathematical treatment of the HITS algorithm are outside the scope of this book but there are some really good resources available online that explain these concepts very well. Some of these resources are also mentioned in the *See also* section of this recipe.

There's more...

Since its introduction, there has been a plethora of research on applications of the HITS algorithm to real-world problems such as finding pages with valuable information on the World Wide Web, a problem otherwise known as webpage ranking. There also has been intensive research on improving the time complexity of the HITS algorithm. A simple search on `http://scholar.google.com/` for HITS will reveal some of the interesting research that has been, and is being, carried out in this domain.

See also

- The Wikipedia page on the HITS algorithm at `http://en.wikipedia.org/wiki/HITS_algorithm` to know more about the HITS algorithm.

- Some great explanation along with real-world examples in the lecture notes by Raluca Tanase and Remus Radu at `http://www.math.cornell.edu/~mec/Winter2009/RalucaRemus/Lecture4/lecture4.html`.

- *Authoritative Sources in a Hyperlinked Environment* by Jon M. Kleinberg that was published in Journal of the ACM at `http://www.cs.cornell.edu/home/kleinber/auth.pdf`. The algorithm used in Gephi for computing the values of hubs and authorities is from this paper.

- Another paper by Jon Kleinberg on the topic titled *Hubs, authorities, and communities* at `http://dl.acm.org/citation.cfm?id=345982`.

- The book titled *Introduction to Information Retrieval* by Christopher D. Manning, Prabhakar Raghavan and Hinrich Schutze.

- `http://www.dei.unipd.it/~pretto/cocoon/hits_convergence.pdf` to know the stopping criterion for HITS in detail.

Finding a graph's modularity

The modularity of a graph is a measure of its strength and describes how easily the graph could be disintegrated into communities, modules, or clusters. In this recipe, the concept of modularity, along with its implementation in Gephi, is described.

How to do it...

To obtain the modularity score for a graph, follow these steps:

1. Load the Les Misérables graph in Gephi.

2. In the **Network Overview** tab under the **Statistics** panel, hit the **Run** button adjacent to **Modularity**.

3. In the **Modularity settings** window, enter a resolution in the textbox depending on whether you want a small or large number of communities:

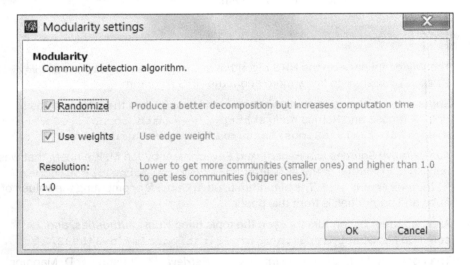

You can choose to randomize to get a better decomposition of the graph into communities, but this increases the computation time.

You can also choose to include edge weight in computing modularity.

4. Hit **OK** once done.

5. This opens up the **Modularity Report** window, which the size distribution of communities into various modularity classes. The report also shows the number of communities formed, along with the overall modularity score of the graph:

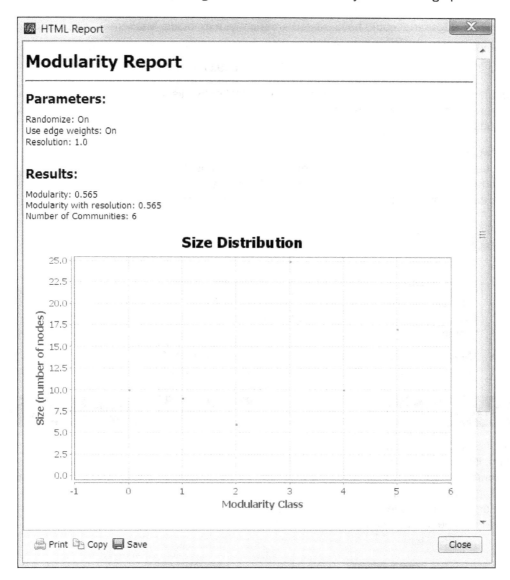

How it works...

Modularity is defined as the fraction of edges that fall within the given modules to the total number of edges that could have existed among these modules. Mathematically, modularity is computed as M = $\sum_{i=1}^{k}\left(x_i - y_i^2\right)$, where x_i is the probability that an edge is in module *i* and y_i^2 is the probability that a random edge would fall into the module *i*.

Modularity is a measure of the structure of graphical networks. It determines the strength of the network as a whole. It describes how easily a network could be clustered into communities or modules.

A network with high modularity points to strong relationships within the same communities but weaker relationship across different communities. It is one of the fundamental methods used during community detection in graphs. Modularity finds its applications in a wide range of areas such as social networks, biological networks, and collaboration networks.

See also

- The Wikipedia page `http://en.wikipedia.org/wiki/Modularity_` `(networks)` to know more about modularity
- The paper titled *Community detection in graphs* by Santo Fortunato at `http://` `arxiv.org/abs/0906.0612` to get an insight into the problem of detecting communities in graphs
- *Modularity and community structure in networks* by M.E.J. Newman (`http://www.` `pnas.org/content/103/23/8577`) is another paper on communities in graphs

Finding a graph's PageRank

Just like the HITS algorithm, the PageRank algorithm is a ranking algorithm for the nodes in a graph. It was developed by the founders of Google, Larry Page and Sergey Brin, while they were at Stanford. Later on, Google used this algorithm for ranking webpages in their search results. The PageRank algorithm works on the assumption that a node that receives more links is likely to be an important node in the network. This recipe explains what PageRank actually is and how Gephi could be used to readily compute the PageRank of nodes in a graph.

How to do it...

The following steps describe the process of finding the PageRank of a graph by making use of the capabilities offered by Gephi:

1. Load the directed version of the Les Misérables network into Gephi.

2. In the **Settings** panel, under the **Network Overview** tab, click on the **Run** button placed against **PageRank**.

3. This opens up the **PageRank settings** window as shown in the following screenshot:

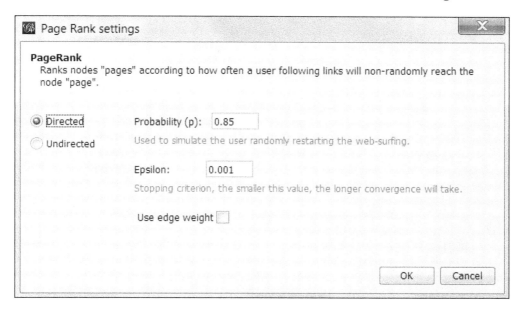

4. Choose which version, **Directed** or **UnDirected**, you want to use for computing the PageRank.

5. In the **Probability** textbox, enter the initial probability value that would serve as the starting PageRank for each of the nodes in the graph.

6. Enter the stopping criterion value in the **Epsilon** textbox. The smaller the value of the stopping criterion, the longer the PageRank algorithm will take to converge.

7. You can choose to include or leave out the edge weight from the computation of the PageRank.

8. Hit **OK** once done.

9. This opens up a new window titled **PageRank Report** depicting the distribution of the PageRank score over a graph.

The following screenshot shows the distribution of PageRank in the directed Les Misérables network with the initial probability as **0.85** and the epsilon value as **0.001**:

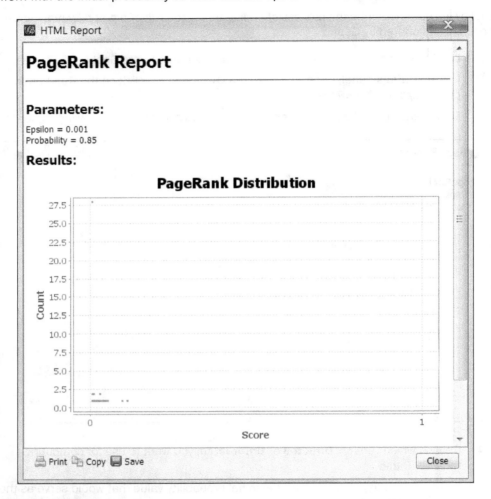

How it works...

The PageRank algorithm, like the HITS algorithm, is a link analysis algorithm and aims to rank the nodes of a graph according to their importance in the network. The PageRank for a node is a measure of the likelihood of arriving at this node starting from any other node in the network through non-random graph traversal.

The PageRank algorithm has found its applications in a wide range of areas including social network analysis, webpage ranking on World Wide Web, search engine optimization, biological networks, chemistry, and so on.

See also

- ▶ *The Anatomy of a Large-Scale Hypertextual Web Search Engine* by Sergey Brin and Lawrence Page, published in the Proceedings of the seventh International Conference on the World Wide Web, which describes the algorithm that is used by Gephi to compute the PageRank. The paper can be downloaded from `http://www.computing.dcu.ie/~gjones/Teaching/CA437/showDoc.pdf`.

Finding connected components in a graph

Connected components in a graph refer to a set of vertices that are connected to each other by direct or indirect paths. In other words, a set of vertices in a graph is a connected component if every node in the graph can be reached from every other node in the graph. In this recipe, you will learn about connected components and how you can run the algorithm to find connected components in Gephi on the graph of your choice.

How to do it...

The following steps illustrate the process to find the connected components in a graph:

1. Load the directed version of the Les Misérables network into Gephi.

2. Click on the **Run** button placed adjacent to **Connected Components** in the **Network Overview** tab of the **Statistics** panel. This opens up the **Connected Components settings** window.

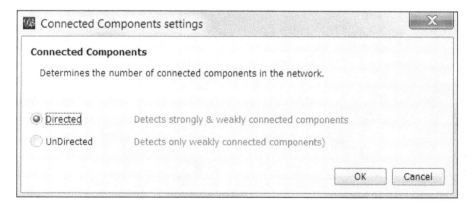

3. Choose between the **Directed** and **UnDirected** versions of the graph. Running the connected components detection algorithm on undirected graphs will lead to detection of only the weakly connected components, whereas running it on directed graphs will lead to detection of both strongly and weakly connected components. The concept of weakly and strongly connected components is explained in *There's more...* section of this recipe.

4. Hit **OK**. A new window titled **Connected Components Report** displaying the size distribution of different components in the chosen graph will be opened up on top of the Gephi application window:

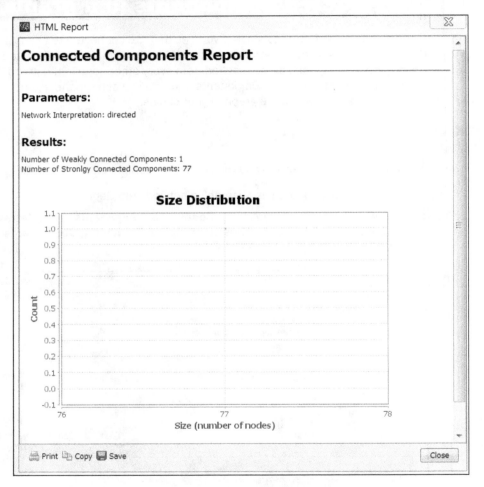

How it works...

The problem of finding the connected components in a graph is easily solved by using simple search algorithms. Beginning with one vertex and then attempting to reach every other vertex in the graph reveals whether it is possible to reach every other vertex in the graph from a vertex, thereby confirming whether the graph is connected or not, and that is the crux of the connected components discovery algorithm. The search can be made by using either the breadth-first search algorithm, the depth-first search algorithm, or any other similar search algorithm. The algorithm implemented in Gephi makes use of the depth-first search algorithm.

There's more...

The connected components could either be weakly connected or strongly connected components. A weakly connected component is one where there is an undirected path between every pair of nodes in the component—that is, every node in the graph is accessible from every other node via an undirected path. A strongly connected component in a subgraph of a directed graph such that every node is accessible from every other node via a directed path in this subgraph.

See also

- The book titled *Introduction to Algorithms* by Thomas H. Cormen, Charles E. Leiserson, Ronald L. Rivest, and Clifford Stein to learn about connected components and search algorithms for graphs in detail.

- The paper titled *DEPTH-FIRST SEARCH AND LINEAR GRAPH ALGORITHMS* by Robert Tarjan that was published in *SIAM Journal on Computing* in 1972 to know more about the algorithm used in Gephi for finding connected components in a graph. This paper is accessible at `http://www.csee.wvu.edu/~xinl/library/papers/comp/Tarjan_siam1972.pdf`.

Getting a node overview of a graph

So far, in all the recipes in this chapter, we have learned how to find out the statistical properties of the nodes in a graph, as well as of the overall graph. Gephi also provides metrics that can be run against a graph to understand the statistical properties of its nodes. In this recipe, some of these metrics, the average clustering coefficient and eigenvector centrality to be precise, are covered.

How to do it...

To find the average clustering coefficients and eigenvector centralities for nodes in a graph, follow these steps:

1. Load the undirected version of the Les Misérables network in Gephi.

2. In the **Statistics** panel, under the **Node Overview** tab, click on **Run** against **Avg. Clustering Coefficient**. This opens up the **Clustering Coefficient settings** window. Select which network version—**Directed** or **UnDirected**—you want to use to find the average clustering coefficient:

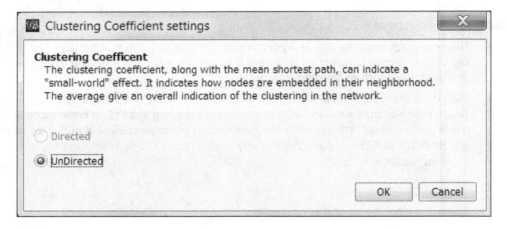

Clustering Coefficient settings X

Clustering Coefficent
The clustering coefficient, along with the mean shortest path, can indicate a "small-world" effect. It indicates how nodes are embedded in their neighborhood. The average give an overall indication of the clustering in the network.

○ Directed

● UnDirected

OK   Cancel

3. Hit **OK**. This opens up the **Clustering Coefficient Metric Report** window, which displays the average clustering coefficient, total triangles, and clustering coefficient distribution:

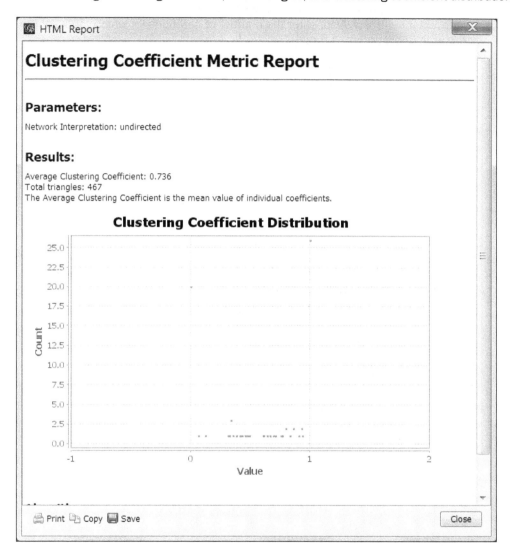

4. In the **Statistics** panel, under the **Node Overview** tab, click on **Run** against **Eigenvector Centrality**. This opens up the **Eigenvector Centrality settings** window. Select which network version—**Directed** or **UnDirected**—you want to use to find the eigenvector centrality. Also, enter the number of iterations for which you want the computation algorithm to run:

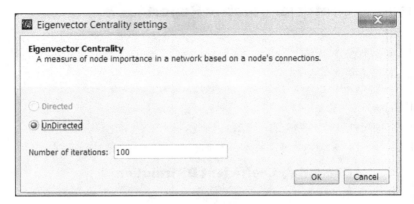

5. Hit **OK**. This opens up the **Eigenvector Centrality Report** window displaying the eigenvector centrality distribution for the graph:

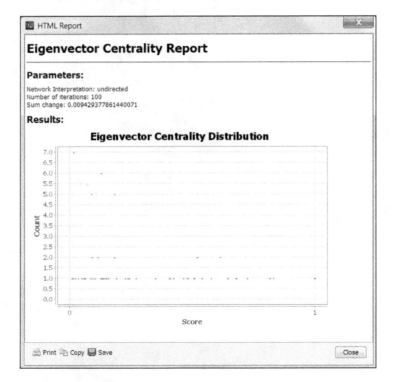

How it works...

The clustering coefficient is a measure of the degree to which the nodes in a graph tend to cluster together. The clustering coefficient of a node *k* is defined as $C = 2e / (n(n-1))$, where *e* is the number of edges between the neighbors of *k* and *n* is the number of number of neighbors of *k*. So, for a node with five neighbors with six edges among them, the clustering coefficient would be *(2 * 6) / (5 * 4) = 0.6*. The clustering coefficient of a node in a graph always lies between zero and one. The average clustering coefficient for a network is then the average of the individual clustering coefficients of its nodes.

In graph theory, the centrality of a node is a measure of the importance of the nodes in the network. There are multiple types of centrality measures that can be defined for a node and eigenvector centrality is one of those. The basic concept behind the eigenvector centrality is that a node with connections to more influential nodes is considered to be more influential than those connected to less influential nodes. As such, the PageRank metric that we covered earlier in the chapter is a type of eigenvector centrality measure.

See also

► The algorithm for computing the average clustering coefficient described in Matthieu Latapy's paper, titled *Main-memory triangle computations for very large (sparse (power-law)) graphs* published in 2008 at `http://www.sciencedirect.com/science/article/pii/S0304397508005392`

► The paper titled *The Anatomy of the Facebook Social Graph* by Johan Ugander, Brian Karrer, Lars Backstrom, and Cameron Marlow (`http://arxiv.org/pdf/1111.4503.pdf`) to understand the application of the average clustering coefficient in social networks

► An interesting application of eigenvector centrality in analyzing patterns in fMRI data in the human brain can be found in the paper titled *Eigenvector Centrality Mapping for Analyzing Connectivity Patterns in fMRI Data of the Human Brain* (`http://www.plosone.org/article/info%3Adoi%2F10.1371%2Fjournal.pone.0010232`)

Getting an edge overview of a graph

Similar to getting an overview of the statistical properties of nodes in a graph, Gephi also allows users to get an overview of the statistical properties of edges in a graph. In this recipe, you will learn about one such statistical property called **average path length**.

How to do it...

The following steps describe how to compute the average path length in Gephi:

1. Load the undirected version of the Les Misérables network in Gephi.

2. In the **Statistics** panel, under the **Edge Overview** tab, click on **Run** against **Avg. Path Length**. This opens up the **Graph Distance settings** window:

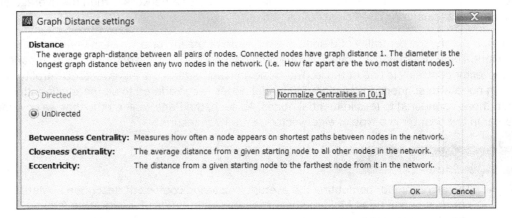

3. In the **Graph Distance settings** window, select the graph type on which you want to find the average path length—**Directed** or **UnDirected**.

4. Check the box next to **Normalize Centralities in [0, 1]** to allow Gephi to normalize the three centrality values between zero and one. The three centralities being referred to here are **Betweenness Centrality**, **Closeness Centrality**, and **Eccentricity**. The concept of betweenness centrality, closeness centrality, and eccentricity has already been covered in the *Finding the network diameter* recipe, earlier in the chapter.

5. Click on **OK**. This opens up the **Graph Distance Report** window, as displayed in the following screenshot, that shows the value of the network diameter, network radius, average path length, number of shortest paths, and three separate graphs depicting betweenness centrality distribution, closeness centrality distribution, and eccentricity distribution:

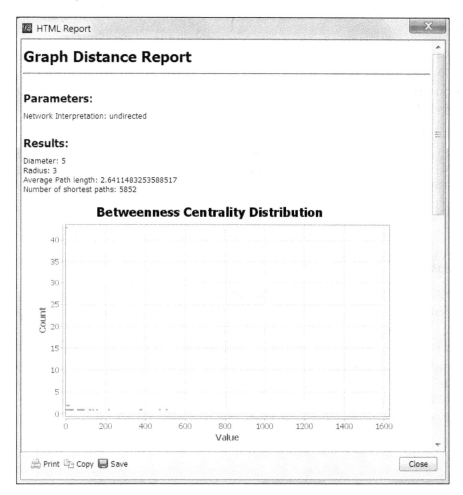

Getting dynamic statistics for a graph

Along with the capabilities available for studying the statistical properties of static graphs, Gephi also offers capabilities for studying the statistical properties of dynamic graphs. In this recipe, we will learn about some of these properties of dynamic graphs and how Gephi helps you study them.

Getting ready

To work on this recipe, we will first need to generate a dynamic graph to work on. To do that, follow these steps:

1. Load Gephi into a blank project mode.
2. Click on the **File** option in the menu bar located on top of the Gephi application window. From the drop-down list, select **Generate** and then **Dynamic Graph Example**. This generates a new dynamic graph in Gephi.

How to do it...

To study the statistical properties of the dynamic graph that we just generated, follow these steps:

1. Click on the **Settings** tab in the **Statistics** panel and select all the options under the **Dynamic** list.
2. Click on **OK**. This populates the **Dynamic** tab of the **Statistics** panel with the following metrics: the number of nodes, number of edges, degree, and clustering coefficient.
3. To see the distribution of nodes over time in this dynamic graph, hit the **Run** button placed adjacent to the number of nodes and enter the sliding window size and the tick information. The sliding window size is the time frame that you want to consider and the tick value is the offset value for this window. Hit **OK**.

This generates the **Dynamic Number of Nodes Report**, as shown in the following screenshot:

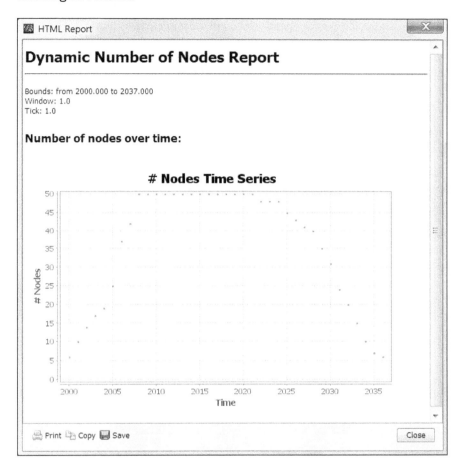

4. To see the distribution of edges over time in this dynamic graph, follow the same steps after hitting the **Run** button adjacent to the number of edges.

5. For the degree of nodes and average degree of graph over the time period considered, hit the **Run** button against **Degree** and then follow the steps you followed in finding the number of nodes.

6. Similarly, for the clustering coefficient, hit **Run** against **Clustering Coefficient**, enter the required information in the **Settings** window and hit **OK**.

Applying individual filters on a graph

Very often, anyone studying a graph is interested in certain parts of it and doesn't need to look at the entire graph, and that is where we need a way of filtering out required portions of the graph. Gephi has a wide set of filters, based on which specific parts of a graph can be filtered out and then studied in detail. For example, one might filter out nodes based on modularity and then run statistics such as degree, PageRank, and so on, one after the other. In this recipe, we will learn about some of these filters and how to use them on graphs.

How to do it...

To filter out selected portions of a graph based on certain pre-specified criteria, follow these steps:

1. Load the Les Misérables network in Gephi.

2. In the **Filters** panel located on the right-side of the Gephi application window next to the **Statistics** panel, you will see a library of filters categorized into buckets based on which part of the graph the filter is going to be applied to:

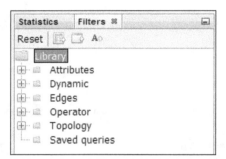

3. Expand one of the buckets, say **Attributes**, and under that select a subbucket, say **Partition**, and then select a criterion for filtering. Here, it is **Modularity Class**.

4. Double-click on **Modularity Class** or drag-and-drop it in the **Queries** panel below the **Filters** panel to select the filter:

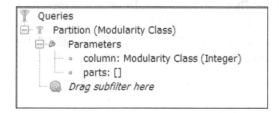

5. In the **Settings** panel situated below the **Queries** panel, click on the square boxes adjacent to the modularity class that you want to show:

6. Hit **Filter** to apply this filter on the graph. The following screenshot shows the Les Misérables network when the **Modularity Class** filter has been applied on its partitions, and only partitions corresponding to modularity classes 7, 6, and 2 have been retained:

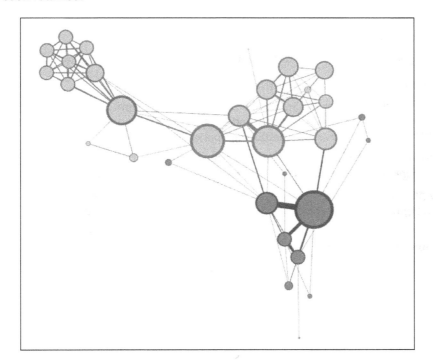

How it works...

On applying a specific filter, Gephi extracts only the node and edge information corresponding to selections made in the filter settings. With this new information, it constructs a new graph and displays it on the screen. This feature comes in handy when one is dealing with large networks and wants to clearly visualize only parts of interest in the graph under consideration.

Apart from **Modularity Class**, there are a lot of other interesting filters available in Gephi, such as edge weight, filtering based on node labels, filtering based on edge labels, filtering based on ID, filtering based on degree range, and so on.

Applying a combination of filters on a graph

In the previous recipe, we learned how to apply individual filters on graphs. In this recipe, we will learn how to apply a combination of filters. So let's get started.

How to do it...

The following steps illustrate how to apply a combination of filters to a graph in Gephi.

1. Load the Les Misérables network in Gephi.

2. In the **Filters** panel located on the right-side of the Gephi application window, next to the **Statistics** panel, expand the **Operator** option and drag-and-drop the **INTERSECTION** filter into the **Queries** panel below.

3. Now, in the **Filters** panel, expand the **Topology** option and drag-and-drop the **Degree Range** filter under the **INTERSECTION** filter in the **Queries** panel.

4. Again, in the **Filters** panel, expand the **Edges** option and drag-and-drop the **Edge Weight** filter under the **INTERSECTION** filter in the **Queries** panel.

5. Double-click on **Modularity Class**, or drag-and-drop it in the **Queries** panel below the **Filters** panel, to select the filter:

6. Click on **Degree Range** in the **Queries** panel to open the **Settings** panel for **Degree Range**. Move the sliders to select the range of degrees to be filtered:

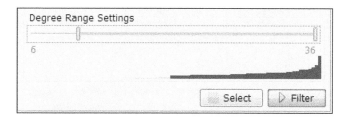

7. Click on **Edge Weight** in the **Queries** panel to open the **Settings** panel for **Edge Weight**. Move the sliders to select the range of edge weights to be filtered:

8. Hit **Filter** to apply this combination of degree range and edge weight filters to the graph. The following screenshot shows the Les Misérables network when only its edges having a weight between **3.61** and **18.1**, and nodes having degrees between **6** and **36**, have been retained:

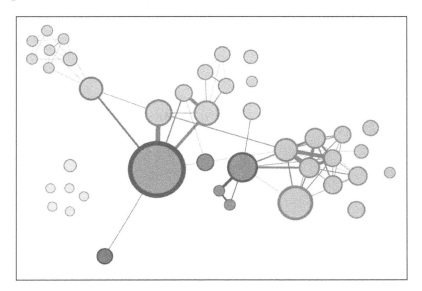

How it works...

When a logical operator is selected, followed by one or more filters, Gephi selects edges and nodes after applying all the filters on all of them and retains only the ones matching the criteria.

There's more...

The results of applying a combination of filters on nodes or edges, based on what sort of filter has been selected, can be exported as columns to be used later on. To do so, click on the first button next to the **Reset** button, which is located at the top of the **Filters** panel. When the mouse pointer is hovered on it, the pop-up text should read, **Export filtered graph as a true/false data column**. In the pop-up box, enter the title of the column that will contain the result of this filtering and hit **OK**.

Filtering dynamic graphs based on time intervals

Dynamic graphs occur almost everywhere in our day-to-day lives and hence it becomes very important for a good visualization tool to allow capabilities for studying these graphs. Gephi has some brilliant features for studying various properties of dynamic graphs. In this recipe, we will learn how to filter dynamic graphs based on varying timelines, so that one can see the status of the network in a time range. Let's get started then.

Getting ready

To work on this recipe, we will first need to generate a dynamic graph to work on. To do that, follow these steps:

1. Load Gephi into a blank project mode.

2. Click on the **File** option in the menu bar that is located at the top of the Gephi application window. From the drop-down list, select **Generate** and then **Dynamic Graph Example**. This generates a new dynamic graph in Gephi.

3. Run the Fruchterman Reingold algorithm on this graph.

4. In the **Partition** panel, under the **Nodes** tab, click on the refresh button to populate the drop-down menu with the **Score** option. Select **Score** and hit **Apply**.

How to do it...

To filter out the nodes and edges of the dynamic graph that we just generated based on varying time intervals, follow these steps:

1. In the **Filters** panel located on the right-side of the Gephi application window, next to the **Statistics** panel, expand the **Dynamic** option and drag-and-drop the **Time Interval** filter into the **Queries** panel below.

2. Click on the **Open Timeline** button in the **Dynamic Range settings** panel that is located at the bottom-right corner of the Gephi application screen:

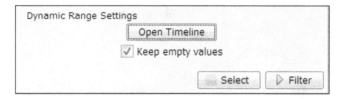

3. Click on **Enable Timeline** in the timeline bar situated at the bottom of the screen.

4. Go to the rightmost end of the timeline until you see a double-ended arrow. Hold and drag the end to change the selected timeframe. You can do it from the beginning of the timeline bar.

5. Now, drag the selected frame across the timeline and notice the change in the edges and nodes of the graph.

The following screenshot shows the dynamic graph that we have created in this recipe with its edges and nodes filtered out for a specific time period:

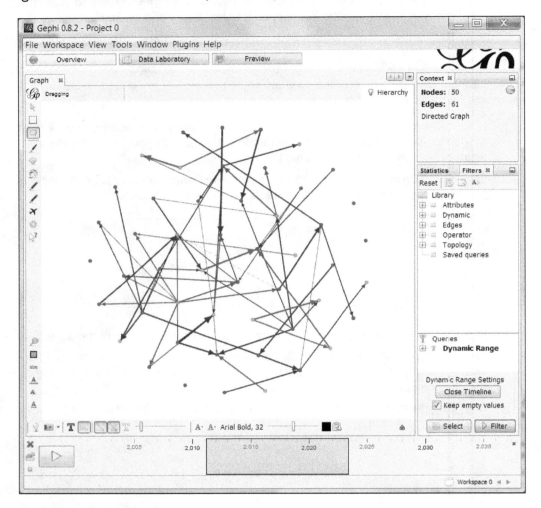

See also

▶ *Chapter 8, Exploring Dynamic and Multilevel Graphs,* for more detail on dynamic graphs

6

Working in the Data
Laboratory Mode

In this chapter, we will cover the following recipes:

- ▸ Importing a spreadsheet
- ▸ Adding and deleting nodes in a graph
- ▸ Changing the attributes of an existing node in a graph
- ▸ Adding and deleting edges in a graph
- ▸ Changing the attributes of an existing edge in a graph
- ▸ Adding/deleting columns
- ▸ Merging columns
- ▸ Copying data between columns
- ▸ Filtering, searching, and modifying data based on particular attributes
- ▸ Creating columns with dynamic regular expression filtering
- ▸ Exporting a table

Introduction

The ability to manipulate data with a good user experience is a key feature for any network analysis and visualization tool. The tool should give its user the flexibility to modify the network data easily, be it nodes and their attributes, edges and their attributes, and so on. Gephi provides its users with a very simple and easy to use interface with which to manipulate data in a tabular form. In this chapter, we are going to learn about the **Data Laboratory** mode in Gephi.

In the **Data Laboratory** mode, the user can perform a range of tasks including importing data in tabular format, adding new nodes and edges, deleting existing nodes and edges, modifying the attributes of existing nodes and edges, adding new columns to the network data, and so on. So let's get started and learn how to accomplish these tasks.

Importing a spreadsheet

The first step in the **Data Laboratory** mode is obviously to get hold of some data. In Gephi, you can upload network data in the form of tabular data, in other words, as a spreadsheet.

Getting ready

In order to get started with this recipe, we will need some tabular data to work with. For this purpose, we will use the Hero Social Network Data. To download the dataset, visit `http://exposedata.com/marvel` and click on the **Hero Social Network Data (csv)** link. Open the dataset file and insert a new row in the beginning of the file. In the first column of this row, insert the text `Source` and insert the text `Target` in the second column. These labels will act as headers while reading the file.

The following screenshot shows this dataset in Microsoft Excel 2013:

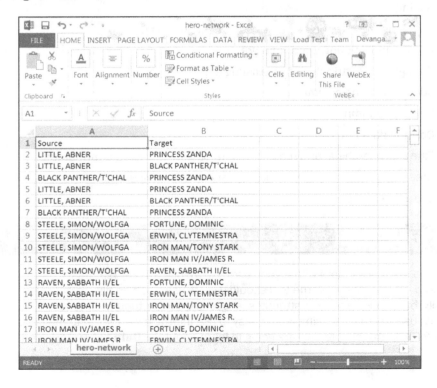

How to do it...

To load the Hero Social Network Data in Gephi, follow these steps:

1. Start Gephi.

2. In the **Welcome** window, select **New Project**.

3. Click on the **Data Laboratory** tab placed in the upper-left side of the Gephi application window:

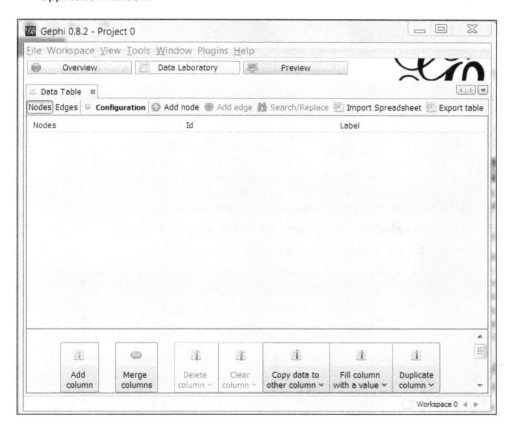

4. Click on the **Import Spreadsheet** button. This opens up the **Import spreadsheet** window. Import the downloaded data file here:

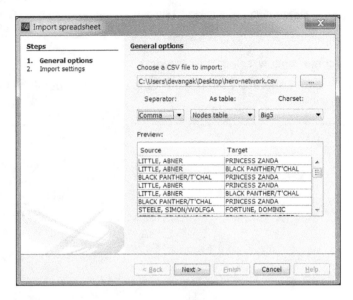

5. In the **As table** drop-down menu, select **Edges table**.

6. In the **Separator** drop-down menu, select **Comma**. Click on **Next**.

7. Tick the **Create missing nodes** checkbox:

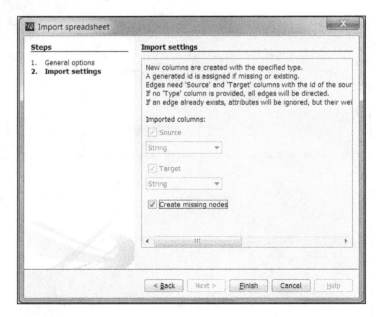

8. Click on **Finish** to complete the import process.

9. This populates nodes and the edges table.

The following screenshot shows how the node data looks. The first column contains the node information, the second column their IDs, and so on:

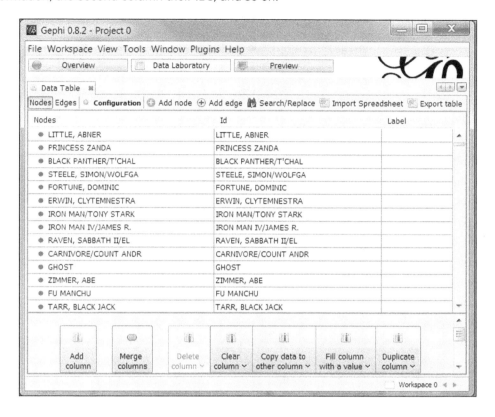

The following screenshot shows the edge data. The first column holds the source nodes, the second column holds the target nodes, and the next column contains the information that tells you whether the link between a particular source and a destination column is directed or undirected, and so on:

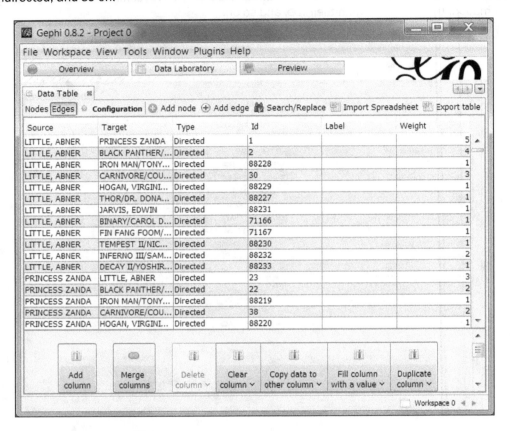

How it works...

The data spreadsheet that has to be loaded in Gephi can either be a node table or an edge table. A node table contains the data about the nodes in the graph, whereas an edge table contains the relationship information. Out of necessity, the edge table has to have two columns—one source column and one target column. Thus, a row entry in an edge table represents an edge that will have the source node from the source column and the destination node from the target column. The edge, depending on the user's choice, can be undirected or directed.

The Hero Social Network Data that we are using for this recipe connects the characters from Marvel comics with each other, depending on which comics that they have both occurred in. This means that two characters will share a relationship if they have appeared in the same comic. These connections are weighted according to the number of their common appearances in a comic.

Adding and deleting nodes in a graph

Manipulating the nodes in a graph is one of the fundamental operations of a graph. One such operation is adding a new node to the existing graph. In this recipe, we will learn how to carry out this operation in Gephi.

Getting ready

In order to understand how to add a new node to the existing graph, we will start with an empty new project and then progressively add nodes to the new network. So, start Gephi and load a new project from the welcome screen.

How to do it...

To add a new node to a new or an existing network, follow these steps:

1. Go to the **Data Laboratory** mode in Gephi.

2. Click on the **Add Node** button in the top panel. This opens the **Add node** dialog window, as shown in the following screenshot:

3. In the **Label** textbox, enter the name that you want to give to the new node. Once entered, hit **OK**. This creates a new node in the network with a user-specified label. The following screenshot shows the spreadsheet view of the network in Gephi with the new node labeled as **Node_1**:

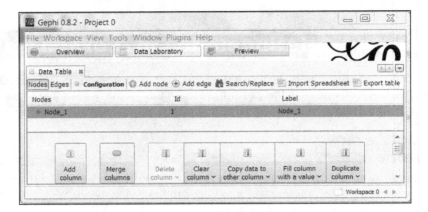

The process for deleting node(s) from a graph is fairly straightforward. Here are the steps to accomplish this:

1. Right-click on the node to be deleted to open up a menu, as shown in this screenshot:

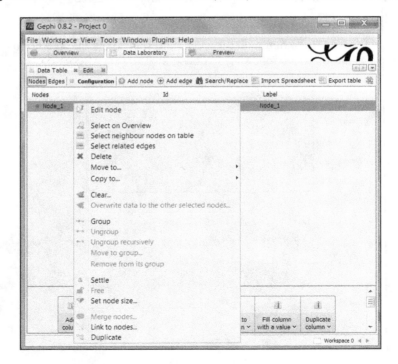

2. Click on **Delete**. This opens up a confirmation window:

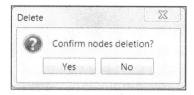

3. Click on **Yes** to delete the node.

Changing the attributes of an existing node in a graph

We often encounter the need to modify the attributes and/or properties of a node or several nodes while working with networks. These attributes/properties include the position of the node in the three-dimensional space, the node label, the node color, and so on. This recipe describes how to accomplish this task.

Getting ready

Open a new or existing project in Gephi. If you have opened a new project, create a new node by following the steps described in the previous recipe. Once the project is loaded, go to the **Data Laboratory** mode.

How to do it...

The following steps describe how to change the attributes of an existing node in a graph:

1. Right-click on the node to open up a menu.

2. Click on **Edit Node**. This opens up a new **Edit** tab, as shown in the following screenshot:

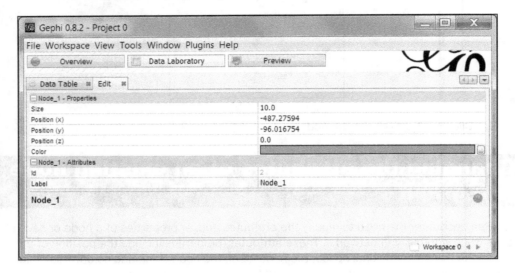

3. Click on the textbox against the property or attribute that you want to change. Enter the new value for the property/attribute and hit *Enter*. This modifies the required property/attribute for the node.

In the following screenshot, the node that we created in this recipe has been relabeled from **Node_1** to **NewNode**:

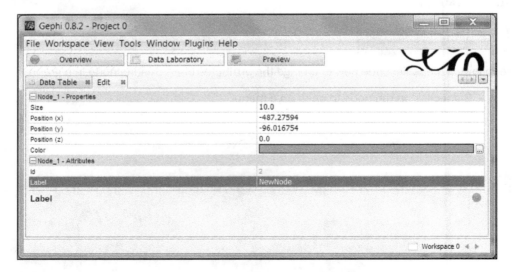

Adding and deleting edges in a graph

Apart from adding nodes to, and deleting them from, the graph, the other operation that is often carried out is the addition and deletion of edges. This recipe teaches you how to carry out this operation in Gephi.

Getting ready

An edge exists between two nodes and hence, before starting to add an edge(s), you will need to make sure that the nodes for this edge exist. So, before we move forward with the recipe, open a new project in Gephi and then, in the **Data Laboratory** mode, add two nodes: **Node_1** and **Node_2**.

How to do it...

With two nodes, **Node_1** and **Node_2**, already existing in our network, follow these steps to add a new edge between these two nodes:

1. Click on the **Edges** button in the top panel.

2. Click on the **Add edge** button in the top panel. This opens a new window, as shown in the following screenshot:

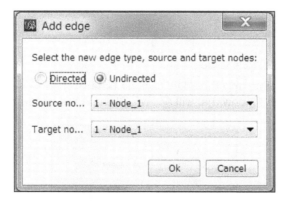

3. Choose between **Directed** and **Undirected**, depending on which type of edge you want to add.

4. Select the source node and target node.

5. Hit **Ok** once this has been done.

Similar to node deletion, the process of deleting an edge from a network or a graph is a fairly straightforward process. However, there are two different ways in which an edge can be deleted. One can choose between deleting only an edge or deleting an edge and its source and target node. These two tasks are covered in the following steps:

1. Click on the **Edges** button in the top panel. This opens up a menu, as shown in the following screenshot:

2. Right-click on the edge to be deleted.

3. If you wish to delete only the edge, click on **Delete**. The following screenshot shows the confirmation window that will appear. Click on **Yes** to delete the edge:

4. If you wish to delete the edge and its source and target nodes, click on **Delete edge with nodes** and click on **Ok**. This opens up a confirmation dialog where you can choose to delete the source node, the target node, or both. Hit **Yes** when this has been done:

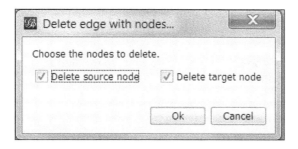

Note that when the source and/or nodes are deleted along with the edge, the other edges incident on these nodes will also get deleted automatically.

How it works...

If the **Undirected** option is selected, the recipe results in the creation of an undirected edge between the two nodes specified. If the **Directed** option is selected, a new edge with its starting node as source node and ending node as target node will be created.

There's more...

A new edge can also be added in the **Overview** mode. The recipe for this has already been discussed in the *Adding nodes and edges to the graph* recipe in *Chapter 2, Basic Graph Manipulations*.

Changing the attributes of an existing edge in a graph

During the network analysis, there is often a need to modify the edges that are already present in the graph. The changes can be with respect to the attributes and properties of the edges such as edge label, color, and so on. In this recipe, we will learn how to edit the existing edges in the **Data Laboratory** mode.

How to do it...

The following steps describe how to change the attributes—such as color, label, weight, and so on—of an existing edge:

1. Click on the **Edges** button in the top panel.

2. Right-click on the edge to be edited to open up the menu list.

3. In the menu, click on **Edit edge**. This opens up the **Edit** tab, as shown in the following screenshot:

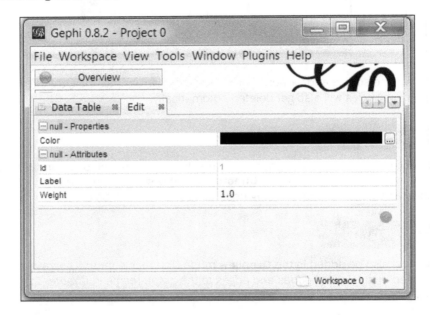

4. Click on the textbox against the property/attribute that you want to edit and enter the new value for that property/attribute. Hit *Enter* once done.

Adding/deleting columns

In previous recipes, we have learned how to carry out the basic operations of adding, deleting, and modifying nodes and/or edges. From this recipe, we will ramp up the complexity level and now deal with columns instead of individual recipes. To begin with, in this recipe we will learn how to add or delete columns in a network in Gephi's **Data Laboratory** mode. So, let's get started.

Getting ready

Before moving forward, download the Hero Social Network Data from `exposedata.com/marvel` by clicking on the **Hero Social Network Data (csv)** link. Open the dataset file and insert a new row in the beginning of the file. In the first column of this row, insert the text as `Source` and insert the text as `Target` in the second column. These labels will act as headers while reading the file. Now, read this file into Gephi using the steps described in the first recipe of this chapter, *Importing a spreadsheet*.

How to do it...

To add a new column to the nodes data, follow these steps:

1. Click on the **Nodes** button in the top panel.

2. At the bottom of the window, you will find a button titled **Add column**, as shown in the following screenshot:

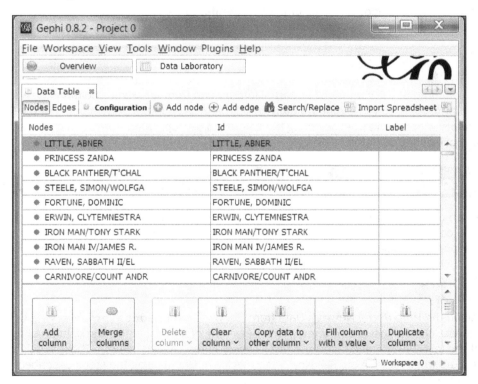

3. Clicking on the **Add Column** button opens up **Add Column – Settings** dialog box, as shown in the following screenshot. Enter the title and select the data type of the new column to be added:

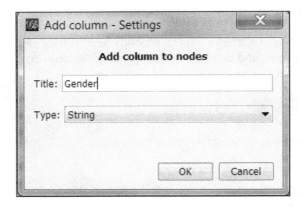

4. Hit **OK**. This adds a new column titled **Gender** to the **Nodes** table. The information in this column will be of type string.

There's more...

A new column that will contain all or partial information from an existing column can also be created by clicking on the **Duplicate column** button present in the bottom panel of the **Data Laboratory** window and then selecting the column to be duplicated from the list of available choices of existing columns.

Merging columns

In certain cases, it is quite possible that the information from two columns can be merged into one column without losing any information about the network and its components. In this recipe, we will learn how to carry out this operation in the **Data Laboratory** mode.

Getting ready

Download the Hero Social Network Data from `http://exposedata.com/marvel`. Open the dataset file and insert a new row in the beginning of the file. In the first column of this row, insert the text `Source` and insert the text `Target` in the second column. These labels will act as headers while reading the file. Now, read this file in Gephi using the steps described in the first recipe of this chapter, *Importing a spreadsheet*.

How to do it...

The following steps illustrate how to merge two columns into a single column in the **Nodes** table:

1. Click on the **Nodes** button in the top panel of the **Data Laboratory** window.

2. Create a new column titled **Gender** in the **Nodes** table by using the previous recipe, *Adding/deleting columns*.

3. Now, click on the **Merge Columns** button located in the bottom panel. This opens up the **Merge columns** dialog window, as shown in the following screenshot:

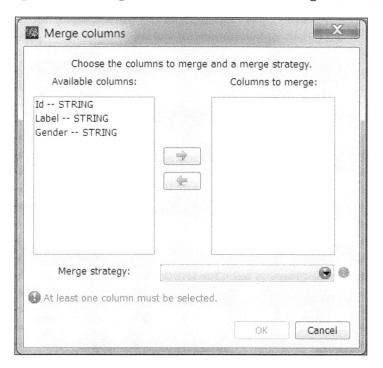

4. Now, move two or more columns that have to be merged from the list of available columns to the **Columns to merge** list by selecting the columns one by one and then clicking on the rightwards pointing arrow. The following screenshot shows the scenario in which the **Label** and **Gender** columns have been chosen as the columns to be merged:

5. From the **Merge strategy** drop-down menu, select the way in which you want to merge these selected columns. In this case, the **Label** and **Gender** columns have been chosen to be merged using the **Join values with separator** strategy.

6. Hit **OK** once done.

This will lead to another window opening up, as shown in the following screenshot:

7. In the **New column title** textbox, enter the name of the new column that will be formed by merging the previously selected two columns.

8. In the **Separator text** textbox, enter the separator for the values picked up from the two columns. Here, we have selected the underscore to be the separator.

9. Hit **OK**.

How it works...

When two columns are merged via a merge strategy, values from the two are picked and transformed into a new value for the new column. In our example, where we combined the **Label** and **Gender** columns by using an underscore, the values from the two columns will be picked up, joined by an underscore, and added to the new column. For example, if the **Label** value for a data point is **1** and the **Gender** is **Female**, then the entry for the **Label_Gender** column for the same data point will be **1_Female**.

Copying data between columns

Quite often, there may arise a need when a column corresponding to some attribute of nodes or edges is a transformation of the values of another attribute. In that case, one would like to copy all the values from one attribute column to another attribute column. Gephi provides a very easy way of accomplishing this task.

Getting ready

Download and open the Hero Social Network Data in Gephi using the steps described in the first recipe of this chapter, *Importing a spreadsheet*.

How to do it...

In order to copy the data from one column to another column in the **Data Laboratory** mode in Gephi, follow these steps:

1. Click on the **Nodes** tab in the top panel.

2. Click on the **Copy data to other column** button located in the bottom panel.

3. From the drop-down, choose the column from which you want to copy the data, as shown in the following screenshot:

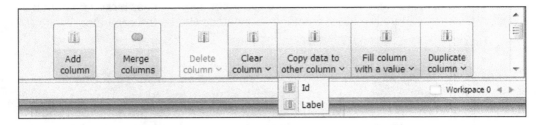

4. In the new dialog window that opens up, choose the column to which the data has to be copied:

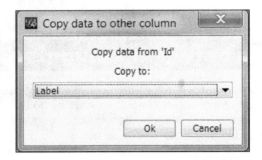

5. Click on **Ok**. This will copy the data from the source column, which is the **Id** column here, to the destination column, which is **Label** in our example.

There's more...

Similar to data copying across columns as described in this recipe, the same task can be accomplished by clicking on the **Duplicate column** button, which is located in the bottom panel, and then giving the name for the new duplicated column in the dialog window. This is shown in the following screenshot:

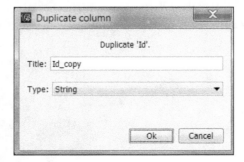

Filtering, searching, and modifying data based on particular attributes

In *Chapter 5*, *Running Metrics, Filters, and Timelines*, we learned how to filter the nodes and edges in a graph, based on certain criteria. Those filters were based on the nodes/edges attributes and properties. The same filtering on node/edge attributes can also be accomplished in the **Data Laboratory** mode. In this recipe, you will learn how to filter the nodes and edges by employing filters on their various attributes.

Getting ready

Download and open the Hero Social Network Data in Gephi using the steps described in the first recipe of this chapter, *Importing a spreadsheet*.

How to do it...

The following steps illustrate the process of filtering out the edges based on their weight:

1. Click on the **Edges** node, located in the top panel.
2. From the drop-down menu next to the **Filter** textbox in the top panel, select **Weight**:

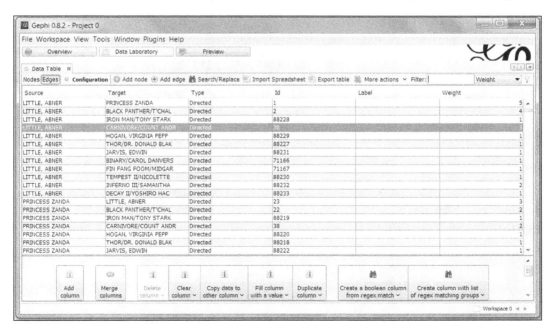

3. In the **Filter** textbox, enter `^5$`, which is the regular expression for selecting weights that are equal to five. This will filter out only those edges that have an edge weight equal to 5.

How it works...

Filtering in the **Data Laboratory** mode requires the usage of regular expressions. For example, if you want to filter out all those edges whose edge weight is equal to 4, then typing `4` in the **Filter** textbox will return back all those edges whose weights are 4, 45, 402, and so on. This means that it tries to do substring matching in this case. So, to filter out only those edges whose edge weight is 4, you have to use the regular expression `^4$`. To filter out edges with weight 4 or 25, the regular expression `^(4|25)$` has to be used.

See also

Regular expressions are extensively used almost everywhere in the programming and computer science world. There are some very good resources available to get a strong understanding of regular expressions:

▶ The book titled *Mastering Regular Expressions* by Jeffrey E. F. Friedl to learn more about regular expressions

▶ A great tutorial titled *Learn Regex The Hard Way* at `http://regex.learncodethehardway.org/book/`

Creating columns with dynamic regular expression filtering

One complex yet elegant operation that often comes in handy while working with large and complex graphs is filtering the graph according to specific criteria. One of the effective ways of doing so is by creating filters for a specific column using dynamic regular expressions. In this recipe, we will learn how to create new columns in the **Data Laboratory** mode by filtering values from existing columns, using filters derived from regular expressions.

Getting ready

Load the Les Misérables network in Gephi directly from the Gephi welcome screen and go to the **Data Laboratory** mode.

How to do it...

In this recipe, we will create a new column for edge data that holds Boolean values, specifying whether or not the edge ID is more than 100. The following steps describe the process of creating this column:

1. Click on the **Edges** tab.

2. Click on **Create a Boolean column from regex match**, which is located at the bottom of the screen.

3. Select **Id** from the drop-down list. This launches a dialog window, as shown in the following screenshot:

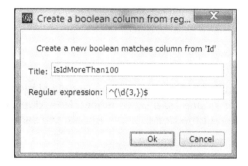

4. Enter the title for the new column and the regular expression that will be used to perform comparisons against the **Id** column.

5. Hit **Ok**. This generates a new column titled **IsIdMoreThan100**, as shown in the following screenshot:

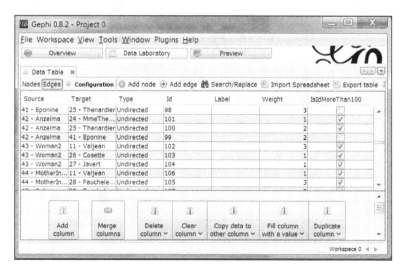

How it works...

The way the Boolean column is generated is pretty simple. The value of the cell from the column selected is matched against the regular expression provided and the Boolean result gets saved in the new column.

Exporting a table

One of the most crucial operations is exporting the network data that you have been using into a reusable format—for instance, into a **comma-separated values** (**csv**) file. This network data could either be completely manually synthesized or it could be a modified version of an already available dataset. In this recipe, you will learn how to export the data into a csv file format.

Getting ready

Download and open the Hero Social Network Data in Gephi using the steps described in the first recipe of this chapter, *Importing a spreadsheet*. Now, duplicate the data from the **Id** column to the **Label** column of the **Nodes** table. Repeat the same thing for the **Label** column of the **Edges** table.

How to do it...

To export network data to CSV format, follow these steps:

1. Click on the **Export table** button located in the top panel of the window. This opens up a dialog box, as shown in the following screenshot:

2. From the **Separator** drop-down menu, select the separator to be used for the data.

3. Select the columns that you want to be exported in the CSV file.

4. Hit **OK**. This opens up the following **Save** dialog box:

5. Enter the desired filename in the **File name** textbox and hit **Save** to save the file.

7

Getting Graphs and Networks Ready for Preview

In this chapter, we will cover the following recipes:

- ▸ Previewing and fine-tuning a graph in the Default mode
- ▸ Previewing and fine-tuning a graph in the Default Curved mode
- ▸ Previewing and fine-tuning a graph in the Default Straight mode
- ▸ Previewing and fine-tuning a graph in the Text outline mode
- ▸ Previewing and fine-tuning a graph in the Black Background mode
- ▸ Previewing and fine-tuning a graph in the Edges Custom Color mode
- ▸ Previewing and fine-tuning a graph in the Tag Cloud mode
- ▸ Exporting a graph in the SVG, PNG, or PDF format

Introduction

So far, we have learned about the various functionalities offered by Gephi that allow one to analyze and manipulate graphs. These functionalities include running various metrics and statistics on graphs, partitioning and ranking nodes and edges, reconstructing graphs using various layouts, adding and deleting nodes and edges, adding new attributes and new properties for nodes and edges, and so on. Hence, being a network analysis and visualization tool, Gephi allows one to export a graph developed into various formats such as PDF, SVG, and PNG after modifying the background, edge color and border, node color and border, and so on. In this chapter, we will cover the rendering settings already available in Gephi and the process of exporting the final graph to multiple formats. So, let's get started.

Previewing and fine-tuning a graph in the Default mode

The **Default** mode in Gephi allows one to export the final graph without performing many manipulations on the nodes and edges. In this recipe, we will learn how to work with the **Default** presets in Gephi's **Preview** mode.

Getting ready

To get started with this recipe, load the Les Misérables network from the **Welcome** screen.

How to do it...

To load and alter default presets in Gephi's **Preview** mode, follow these steps:

1. Click on the **Preview** tab, which is located at the top of Gephi's application window.

2. The **Default** preset should be the one selected by default under the list of presets, as shown in the following screenshot. This populates the nodes and edge attributes with their default values:

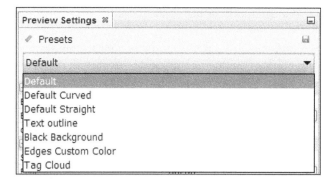

3. Click on the **Refresh** button to load the preview. The following screenshot shows the Les Misérables network when viewed in the **Default** preview mode:

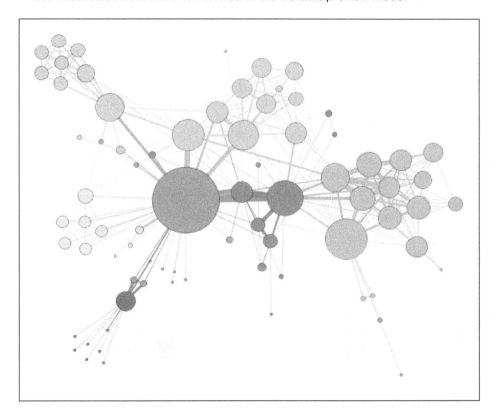

4. You can zoom in or zoom out by using the **zoom** function, which can be found at the bottom of the **Preview** screen.

5. You can also choose to have a background in a color other than white. Just click on the **Background** button that can be found at the bottom of the **Preview** screen and select the color you want to set as the background color.

How it works...

In the **Default** mode, the graph is rendered in almost the same way as defined by the user while building it up. There are minimal customizations possible. This preset works quite well with small graphs where the graph as such is quite clear and is to be saved by the user for further investigation or reference. Large networks might not be rendered as well and may even be totally unusable on exporting. That is when the user might want to look at using more sophisticated rendering mechanisms that can be customized in such a way that the graph is clear and usable. We are going to cover those advanced rendering mechanisms in the upcoming recipes.

There's more...

As mentioned earlier, the default preview doesn't let the user do a lot of fancy decorations with the graph. But there are still a few tweaks that the **Default** preview lets the user carry out in the graph. Here some of them are explained:

► The user can change the nodes' border width, color, and opacity by setting required values in the **Border Width**, **Border Color**, and **Opacity** boxes under **Nodes**

► The user can choose to show node labels by checking the **Show Labels** box

► The user can choose to show/hide edges by checking/unchecking the **Show Edges** box

► The attributes of edges—such as their color, thickness, and so on—can be altered

► The user can choose to show node labels by checking the **Show Labels** box

Previewing and fine-tuning a graph in the Default Curved mode

A very similar rendering mechanism to the **Default** preview is the **Default Curved** preview. The only difference is that by default, the edges in the **Default Curved** preview are curved and the node and edge labels are present for a more aesthetically pleasing and detailed graph. In this recipe, we will learn more about this preview.

Getting ready

To get started with this recipe, load the Les Misérables network from the **Welcome** screen.

How to do it...

The following steps describe how to work with the **Default Curved** rendering in Gephi:

1. Click on the **Preview** tab by default, located at the top of the Gephi application window just below the menu bar.

2. From the **Presets** drop-down list, select the **Default Curved** option.

3. Click on the **Refresh** button to refresh the preview for the graph in the **Preview** window. The following screenshot shows how the Les Misérables graph would appear in the **Default Curved** preview:

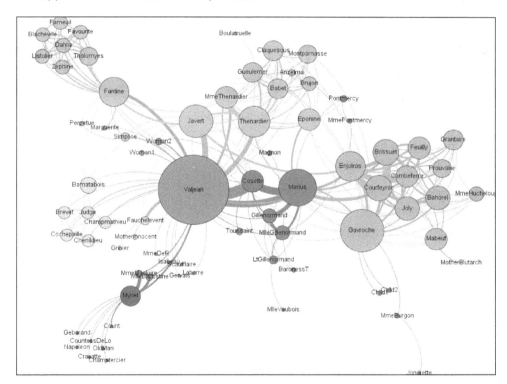

4. If the node labels are unclear and are overlapping each other, you can reduce the node label size by setting an appropriate font in the **Node Labels** options. If you are using international fonts, try switching to **Monospaced** to see the **Unicode** font label. Click on the **Refresh** button once you are done.

How it works...

The **Default Curved** preview mode provides the user with a more aesthetically pleasing version of the graph. The edges are curves instead of straight as in the case of the default mode and the labels, both for nodes as well as edges, are shown, if present. The graph rendered is more detailed as compared to the one rendered in the **Default** mode.

There's more...

Apart from those present in the **Default** mode, the **Default Curved** mode provides additional customizations that can be done to the graph before it is exported. Some of these are discussed as follows:

- ▶ The user can define the display settings for the labels such as the outline size, color, opacity, and so on, both for node labels as well as edge labels
- ▶ The user can define the thickness for the edge arrows in the case of directed graphs

Previewing and fine-tuning a graph in the Default Straight mode

Some of the features available in Gephi are easily derivable from each other and, hence, sometimes they might seem redundant to users. But they often prove to be helpful for beginners. The **Default Straight** mode is one such feature. In this recipe, we will learn more about the **Default Straight** mode.

Getting ready

Load the Les Misérables graph in Gephi from the **Welcome** screen.

How to do it...

The following steps describe how to work with the **Default Straight** rendering mode in Gephi:

1. Go to the **Preview** mode in Gephi.
2. From the **Presets** drop-down list, select the **Default Straight** option.

3. Click on the **Refresh** button to refresh the preview for the graph in the **Preview** window. The following screenshot shows how the Les Misérables graph would appear in the **Default Straight** preview.

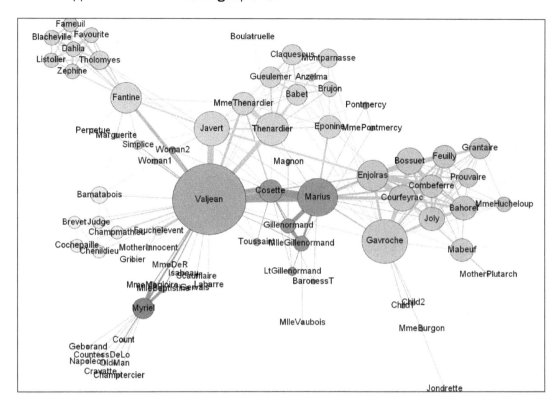

4. If the node labels are unclear and are overlapping each other, you can reduce the node label size by setting an appropriate font in the **Node Labels** options. Click on the **Refresh** button once you are done.

How it works...

The **Default** mode is almost the same as the **Default Curved** preview mode, apart from that in the **Default Straight** mode, a user sees straight edges. Thus, the **Default** mode is more like an amalgamation of the **Default** and **Default Curved** modes.

Previewing and fine-tuning a graph in the Text outline mode

The **Text outline** mode in Gephi renders the graph in such a way that the node and edge label text is surrounded by an outline, thereby placing more emphasis on textual aesthetics.

Getting ready

To get started with this recipe, load the Les Misérables network from the **Welcome** screen.

How to do it...

Here are the steps to render a graph in Gephi in the **Text outline** mode:

1. Click on the **Preview** tab located just below the menu bar in the Gephi application mode.

2. Select **Text outline** from the **Presets** drop-down list in **Preview Settings**, as shown in the following screenshot:

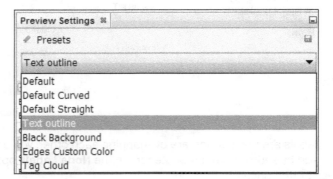

3. Click on the **Refresh** button to reload the rendered graph in the **Text outline** preview mode.

4. If the labels are too big and unclear, uncheck the **Proportional size** box under **Node Labels**. Then click on the box next to the **Font** option and reduce the font size.

The following screenshot shows how the Les Misérables graph looks when rendered in the **Text outline** preview mode and after the label sizes are adjusted for better aesthetics:

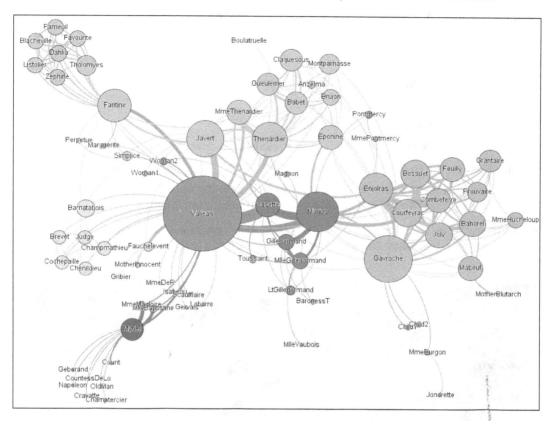

How it works...

By default, the **Text outline** mode produces almost the same graph as the **Default Curved** mode, except that the labels are surrounded by an outline to put more emphasis on the label text and probably make it look less harsh.

Previewing and fine-tuning a graph in the Black Background mode

The **Black Background** mode gives another interesting preview option to the user where the graph is rendered with a black background. Let's look into the details of this mode.

Getting ready

Load the Les Misérables network in Gephi from the **Welcome** screen.

How to do it...

To render the Les Misérables graph with a black background, follow these steps. The steps remain the same for any other graph.

1. Go to the **Preview** mode in Gephi.
2. Select **Black Background** from the **Presets** drop-down list under **Preview Settings**.
3. Click on the **Refresh** button to render the graph.
4. Uncheck the **Proportional size** checkbox and set the appropriate font from the **Font** option under **Node Labels**. Click on the **Refresh** button again.

The following screenshot shows how the Les Misérables graph will look when it has been rendered in the **Black Background** preview mode and the labels have been set for better viewing:

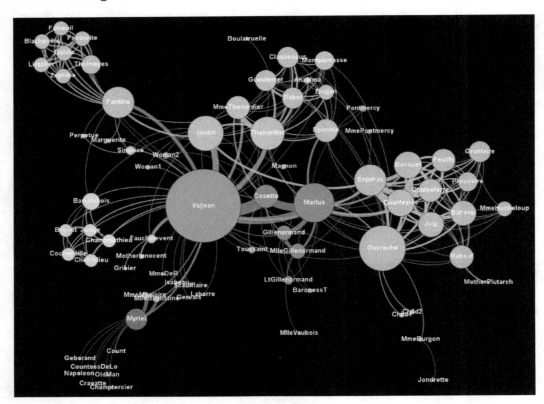

How it works...

This is yet another preview mode that can easily be derived from another mode, in this case the **Default Curved** mode. The only difference is that the graph is rendered on a black background and the colors for nodes, edges, and labels are set accordingly. This mode just gives the user a one-click capability for fixing all of those attributes.

Previewing and fine-tuning a graph in the Edges Custom Color mode

There are cases where the way in which nodes are displayed is less important than the edges themselves. In other words, the connections are more crucial than the entities in the system themselves. The **Edges Custom Color** preview mode comes in handy in such cases. In this recipe, we will learn how differently a graph gets rendered in the **Edges Custom Color** mode, as compared to the other preview modes present in Gephi.

Getting ready

To get started with this recipe, load the Les Misérables network from the **Welcome** screen.

How to do it...

To render the Les Misérables graph in the **Edges Custom Color** preview mode, follow these steps:

1. Go to the **Preview** mode in Gephi by clicking on the **Preview** tab located just below the menu bar in the Gephi application window.
2. From the **Presets** drop-down menu, select **Edges Custom Color**.
3. Click on the **Refresh** button to render the graph in this mode.
4. If the labels do not appear properly or are overlapping each other, uncheck the **Proportional size** checkbox and set the appropriate font size in the **Font** option under **Node Labels**.
5. Click on the **Refresh** button to re-render the graph.

The following screenshot shows how the Les Misérables graph is rendered in the **Edges Custom Color** preview mode with the node labels adjusted for better clarity and rendering:

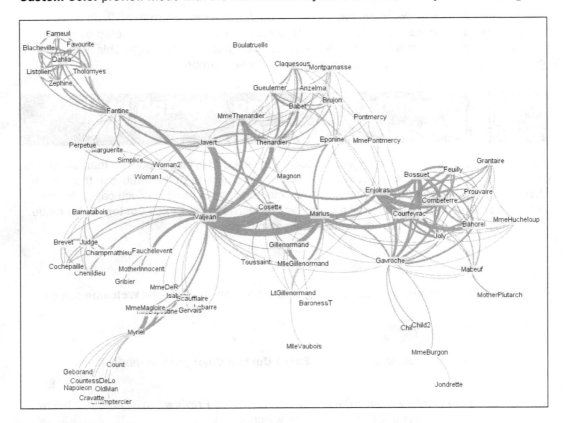

How it works...

In the **Edges Custom Color** preview mode, the graph is rendered in such a way that the nodes disappear and the node labels take their place. The edges are all displayed in the same color by default. The edge weights are preserved and the thickness for the edges is set proportionally.

There's more...

Instead of a single color, the edges can be given other preset colors too. To do so, click on the **Color** option under **Edges** and select **Source** as the option in the **Preview Settings - Color** window. Hit **OK** once done:

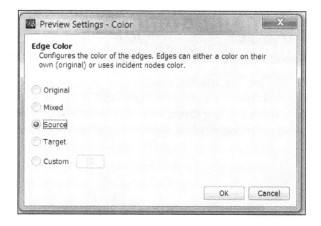

The following screenshot shows the Les Misérables graph when its edges have been configured to take the color of their respective source nodes:

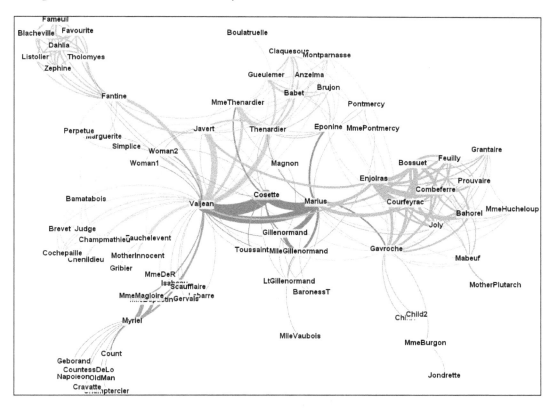

Previewing and fine-tuning a graph in the Tag Cloud mode

One of the most widely used modes, the **Tag Cloud** preview mode, is the one that is most significant in practice. Quite often, the emphasis in network analysis is to find communities of related entities. That is when the **Tag Cloud** mode presents the user with a one-click option to render the graph in such a way that the communities are easily distinguishable and comprehensible.

Getting ready

To get started with this recipe, load the Les Misérables network from the **Welcome** screen.

How to do it...

The following steps demonstrate how to render the Les Misérables graph in the **Tag Cloud** mode. The steps remain the same for any other graph or network as well.

1. Go to the **Preview** mode by clicking on the **Preview** tab located right under the menu bar.

2. From the **Presets** drop-down menu, as shown in the following screenshot, select | **Tag Cloud**:

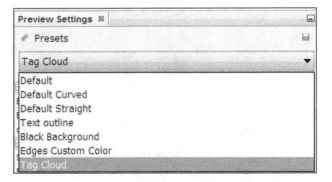

3. Click on the **Refresh** button to render the graph in the **Tag Cloud** preview mode.

4. The graph that appears isn't very appealing. To solve that problem, uncheck the **Proportional size** checkbox and set the appropriate font in the **Font** option under **Node Labels**.

5. Under **Edges**, check the **Show Edges** box.

6. Hit the **Refresh** button again to re-render the graph. You will see the rendered Les Misérables graph, as shown in the following screenshot:

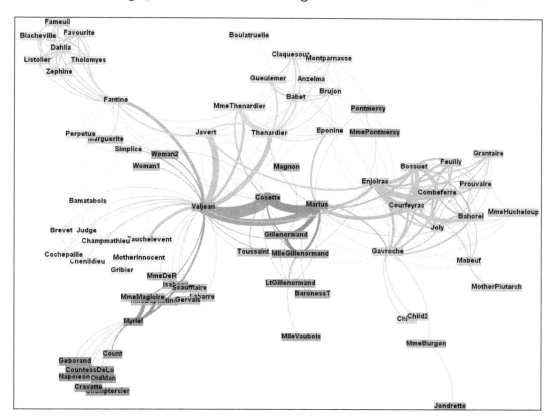

How it works...

In the **Tag Cloud** mode, the nodes disappear and the node labels take their place. The node labels have boxes surrounding them that have the same color as their modularity classes. In other words, labels of the same color belong to the same modularity class. This facilitates the discovery of communities that are essentially entities that have similar attributes and belong to the same class.

Exporting a graph in the SVG, PNG, or PDF format

The next logical step to building up a network—manipulating it based on specific requirements and then rendering it in a clear, aesthetically-pleasing, and practically-usable format—is to export it as a file, either as an image or in a portable document format. In this recipe, you will learn how to carry out the process of exporting a graph in Gephi.

Getting ready

To get started with this recipe, load the Les Misérables network from the **Welcome** screen.

How to do it...

The following steps illustrate how a graph in Gephi can be exported into the SVG, PNG, or PDF formats:

1. Click on the **Preview** tab, which is located just below the menu bar in the Gephi application window, to go the **Preview** mode.

2. Choose one of the preview modes and set different attributes, according to your requirements.

3. Once done, hit the **Refresh** button to render the graph.

4. Change the zoom level by clicking on the buttons marked with a plus and minus sign next to the **Reset Zoom** button, which is located in the bottom-left corner of the **Preview** window.

5. Modify the background, if you want, by clicking on the **Background** button located in the bottom-left corner of the **Preview** window and then selecting the desired color from the window that appears. Hit **OK** once you have chosen the desired background color.

6. Click on the button titled **SVG/PDF/PNG** located in the bottom-left corner of the window right next to the **Export** label:

7. This opens up the **Export** window, as shown in the following screenshot. Give a name to the file in the **File name** textbox. From the **Files of type** drop-down menu, select the file type you want to export the graph to:

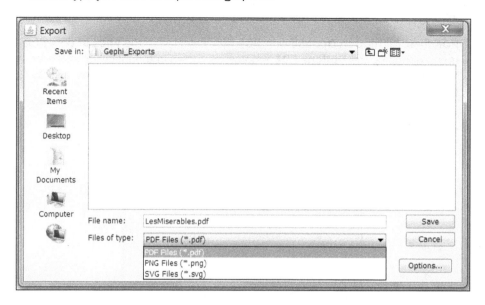

8. Hit **Save** when done.

How it works...

PDF is the abbreviation for **Portable Document Format**. The documents saved in this format are presented independently of the machine hardware, operating system, and application software. PNG refers to **Portable Network Graphics**. It is a file format for images where each image is represented as a dot matrix data structure. PNG supports lossless data compression, which means the images saved in this format are of high quality. SVG refers to the **Scalable Vector Graphics** format that is an XML-based vector image format for two-dimensional images.

There's more...

You can customize the way the required export files are generated. Here are the details about various options that are available while exporting the graph in the three available formats:

> **PDF files**: Clicking on **Options** with **PDF files** selected in the **Files of type** drop-down menu opens up the **Options PDF** window, as shown in the following screenshot. You can set the page size, rendered graph orientation, page margins, and so on in this window:

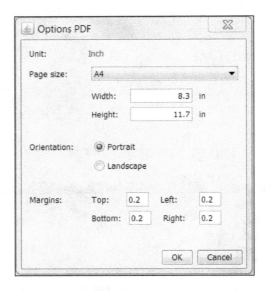

> **PNG files**: When you choose to export the rendered graph as a PNG file, you can specify attributes such as width, height, and margin in the **Options png** window:

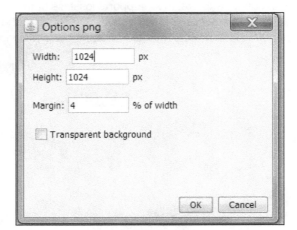

▶ **SVF files**: In the **Options SVG** window that appears when the user chooses to export the graph as a SVG file, the user can specify whether or not to scale stroke width. Stroke width refers to the width of the outline of the graph and is helpful when working with this SVG file in Adobe Illustrator:

Graph files can also be exported to other formats such as CSV, GDF, GEXF, and so on by clicking on **File** under the menu bar, clicking on **Export** then **Graph file,** and finally selecting the format under **Files of type**. This option, however, will let the user download graphical data in a format such that the graph can be loaded again for further manipulation and exploration.

8
Exploring Dynamic and Multilevel Graphs

In this chapter, we will cover the following recipes:

- ▶ Building dynamic/temporal graphs in Gephi
- ▶ Working with dynamic/temporal graphs
- ▶ Working with multilevel graphs
- ▶ Expanding and contracting subgraphs in metanodes
- ▶ Clustering links and attributes

Introduction

So far, in all our recipes, we have dealt with static graphs. Static graphs are those graphs in which the nodes as well as the edges are fixed and do not change over time. In other words, the system that is represented by these graphs changes neither its entities nor the relationships between them over time. Also, the graphs that we've dealt with were all single-level graphs. That is, the relationships present between the entities didn't possess any hierarchies.

In this chapter, we are going to dive into two special kinds of graphs: dynamic graphs and multilevel graphs. These graphs come in handy while representing the real-world systems that keep changing with time and where the relationships among entities are much more complex.

Building dynamic/temporal graphs in Gephi

In this recipe, we will learn about dynamic graphs, otherwise known as **temporal graphs**, and how to build these graphs in Gephi.

How to do it...

The following steps describe how to create a dynamic graph in Gephi:

1. Open Gephi and, on the **Welcome** screen, select **New Project**.

2. Click on **File** in the menu bar. From the drop-down list, select **Generate**, followed by **Dynamic Graph Example**. This is shown in the following screenshot:

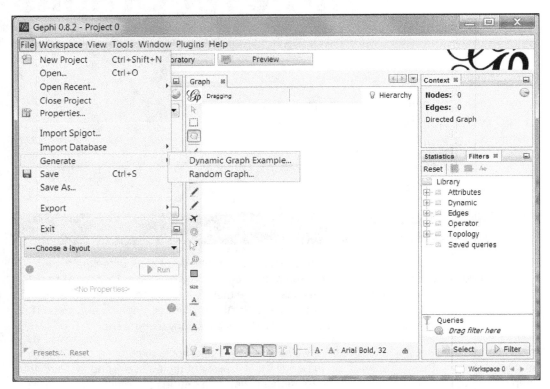

3. Zoom in on the graph for better clarity. You can also run the Fruchterman Reingold Layout algorithm from the **Layout** menu for a better looking graph. The following screenshot shows how the graph will look like after these two operations:

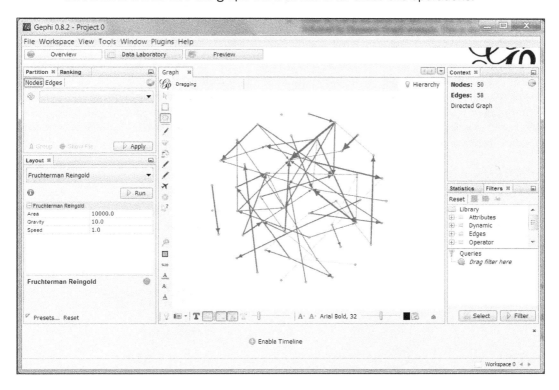

4. Click on **Enable Timeline**, which is located at the bottom of the screen, to view the timeline. This timeline shows that the system being visualized exists between the time interval of 2000 to 2038.

How it works...

Dynamic graphs are a special category of graphs that are used to model dynamic real-world systems. In terms of data, dynamic graphs are those graphs that have time associated with nodes and edges. This means that the system that is modeled as a dynamic graph will have its nodes and edges changing over time. In other words, the entities and the relationships between these entities in the system will constantly be evolving over time. An example of a dynamic system is the employee network within an organization. The employees, as well as the relationships between them, keep changing over time. The relationship between the two employees could involve being employed on the same project. New people join the project and some existing ones leave. This way, the system keeps changing dynamically over time and hence is represented with the help of dynamic graphs.

There's more...

The graph in the **Overview** mode, at first glance, might look like a static graph. However, a look at the graph details in the **Data Laboratory** mode reveals the details and the way they differ from the static graphs.

The following screenshot shows how the node data for the graph that we just generated, looks:

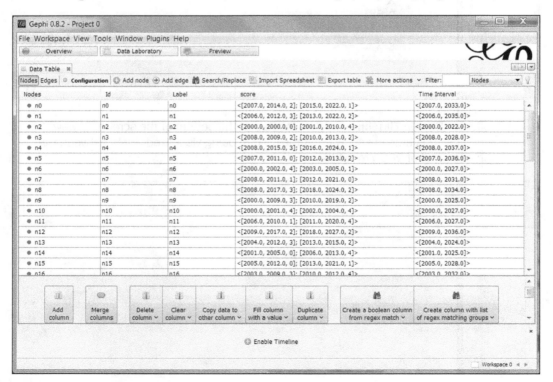

The two columns of interest here are **Time Interval** and **Score**. The **Time Interval** column refers to the life duration of the node. The **Score** column gives the weight of the node at different time intervals. For instance, the node with the **n11** ID is alive only in the duration 2006.0 to 2027.0. This means that this node entered the system at the time instance 2006.0 and stayed until 2027.0. The score of the same node from 2006.0 to 2010.0 is 1 and that from 2011.0 to 2020.0 is 4.

The following screenshot shows the **Edges** data for the same graph in the **Data Laboratory** mode:

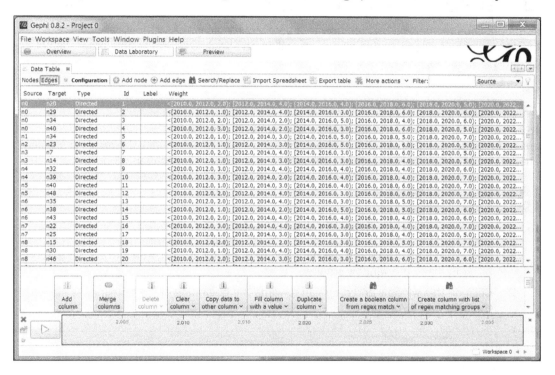

The attribute of interest here is **Weight**. The **Weight** attribute represents the lifetime of the edge and the score of the edge in different time frames within that lifetime. You will notice the tuples in weight have the start time, the end time, and the weight of the edge in that time period.

See also

▸ The paper titled *Dynamic graphs* by Camil Demetrescu, Irene Finocchi, and Giuseppe F. Italiano (http://www.diku.dk/PATH05/CRC-book1.pdf) to know more about dynamic graphs and dynamic graph algorithms

▸ http://www.slideshare.net/Eldarion/gephi-dynamic-features, which is a great tutorial on dynamic graphs in Gephi by Sébastien Heymann

▶ The paper titled *Temporal Graphs* by Vassilis Kostakos at `http://arxiv.org/abs/0807.2357`

▶ The tutorial by Clement Levallois at `http://www.clementlevallois.net/gephi/tuto/gephi_tutorial_dynamics.pdf` for more detail about the creation and manipulation of dynamic graphs

Working with dynamic/temporal graphs

In the previous recipe, you learned how to create a dynamic graph, otherwise known as a temporal graph, in Gephi. In this recipe, you will learn about some interesting properties of dynamic graphs. So, let's get started.

Getting ready

To get started with this recipe, first create a dynamic graph in Gephi as explained in the previous recipe, *Building dynamic/temporal graphs in Gephi*. Also, once the graph is created, enable the timeline that is located at the bottom of the Gephi application window.

How to do it...

To explore dynamic graphs in Gephi, follow these steps:

1. Place the mouse cursor at the right end of the timeline and drag it towards the left in order to make the window smaller.

2. Click on this window and drag it in towards the beginning of the timeline, as shown in the following screenshot:

3. Click on the play button that is located on the left side of the timeline. This will result in the window moving forward towards the right side of the timeline. You will notice that the graph changes accordingly. Some new nodes and edges get added and some existing ones get deleted. Some edges also change their weights over time.

4. You can change the range of the timeline and also define a custom window size. To do so, click on the **Settings** button, which is located on the left side of the timeline, and select **Set Custom Bounds** in the menu that appears.

5. In the **Custom time bounds & interval** window that opens, which is shown in the following screenshot, enter the range for the timeline and the time window size that you want. Click on **OK** once done:

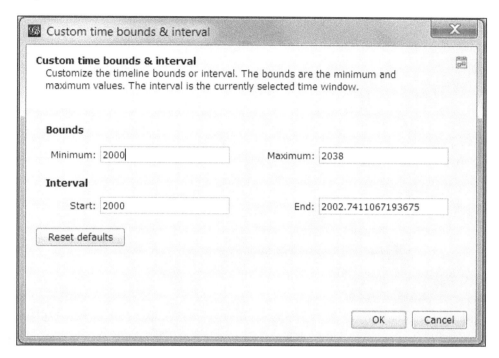

6. You can also change the settings for the time window when the play button is pressed. To do so, click on the **Settings** button that is located on the left side of the timeline window and click on **Set Play Settings**.

7. In the **Animation settings** window, which is as shown in the following screenshot, enter the delay time, step size, mode, and direction for the animation. Hit **OK** when done:

How it works...

When the time window slides over the timeline, it picks up nodes and edges that are alive during that time frame and displays them. These nodes and edges remain visible until they reach their end stage.

There's more...

You can change the way information about the nodes and edges of these dynamic graphs appears in the **Data Laboratory** mode. To do so, follow these steps:

1. Go to the **Data Laboratory** mode and click on the **Nodes** tab.

2. Click on the **Configuration** button to open the **Configuration** window, as shown in the following screenshot:

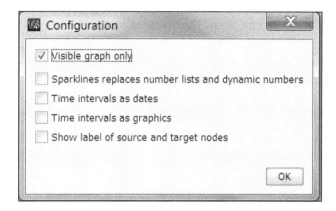

3. Checking the **Sparklines replaces number lists and dynamic numbers** checkbox will make the score information for the nodes appears as shown in the following screenshot:

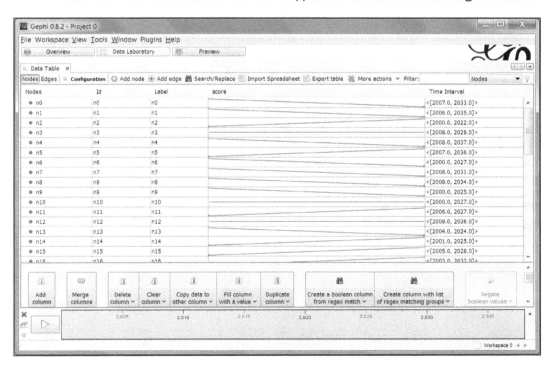

4. You can choose to view the time in a different format by clicking on the **Time intervals as dates** checkbox.

5. The time intervals can also be viewed in graphical format by checking the **Time intervals as graphics** checkbox. The following screenshot shows this:

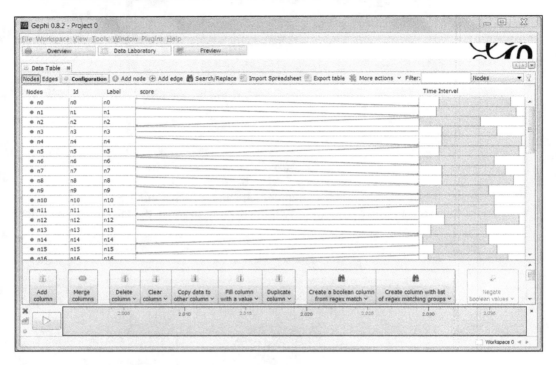

You can also choose to see the source and destination node labels. The same modifications will work for edges too.

See also

▶ The video titled *Temporal Graphs for Dynamic Network Analysis* at `http://videolectures.net/complexnetworks2012_musolesi_temporal_graphs` to get an insight into the usage of dynamic graphs in network analysis

Working with multilevel graphs

So far, whatever graphs and networks we have seen were all single-level networks. This means that the relationships between various nodes aren't defined to be present at multiple levels. In this recipe, we will learn about multilevel graphs and what features Gephi possesses to manipulate and visualize these graphs. We will begin with the single-level Les Misérables graph available for use in Gephi and then modify it to a multilevel graph.

How to do it...

To transform the single level Les Misérables graph to a multilevel graph, follow these steps:

1. Load the Les Misérables graph in Gephi.
2. Click on the **Rectangle Selection** button located on the left side panel of the **Graph** window.
3. Click on the upper-left blank area of the screen and draw to create a rectangle covering a bunch of graph nodes, as depicted in the following screenshot:

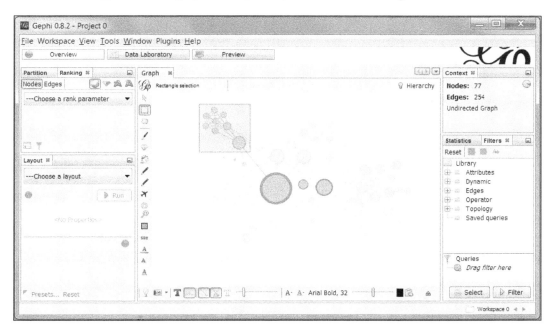

4. Right-click on one of the nodes in the selected area to open up a menu, as shown in the following screenshot:

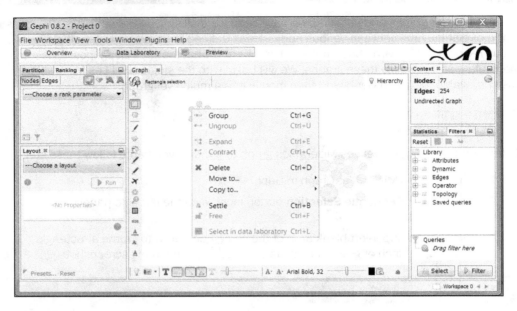

5. Click on **Group** to group these selected nodes together into a single node.

6. Repeat the same process for the other nodes. The graph after a couple of groupings will look something like the one in the following screenshot:

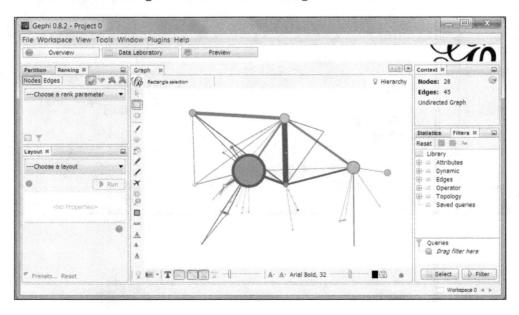

7. You can repeat the same process on already grouped nodes to create another level of hierarchy. To see the levels that are present in the graph, click on **Hierarchy** in the top-right corner of the **Graph** window. A list with the levels and their details pops down on the screen, as can be seen in the following screenshot:

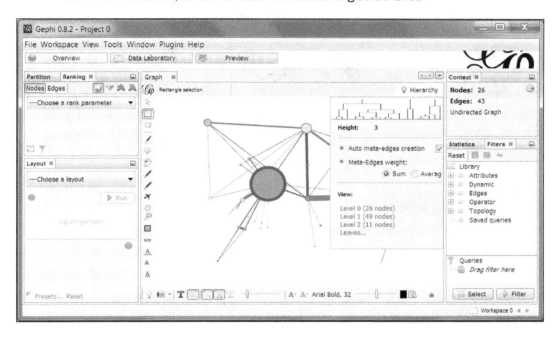

8. Clicking on **Level 2 (11 nodes)** displays the nodes that were grouped in the second level, and so on.

How it works...

Multilevel graphs are very important when it comes to representing real-world data. Many real-world systems possess hierarchy in one form or another. For instance, consider an organization's employee network. The employees are the entities and their presence on the same project or in the same department are the relationships between them. These relationships may not be present at a single level. The CEO of the organization might be present at the top-most level, followed by the department heads on the next level. The department heads may be supervising some project managers individually, who in turn will have other people working under them. This network clearly cannot be efficiently represented at a single level and needs a way of being visualized with all these hierarchies intact. Gephi provides functionalities that allow these visualizations. Multilevel graphs are often also called **hierarchical graphs**.

See also

▶ The paper titled *Hierarchical graph maps* by James Abello at `http://www.mgvis.com/Papers/HierGraphMap_sdarticle.pdf` to know more about hierarchical graphs and an interesting implementation of these graphs

Expanding and contracting subgraphs in metanodes

In the previous recipe, you learned how to select nodes in graphs and group them together to build up hierarchical relationships. In this recipe, we will learn more about the grouping of related nodes into a node referred to as the metanode and degrouping/expanding nodes from the metanode.

Getting ready

To get started with this recipe, first open the Les Misérables graph in Gephi.

How to do it...

The following steps illustrate how nodes in Gephi can be grouped as metanodes and conversely, how metanodes can be expanded into their individual nodes:

1. Group together a bunch of nodes using the previous recipe, *Working with multilevel graphs*.

2. Repeat this process multiple times for different groups.

3. To expand the group, right-click on the node that represents the group, called the metanode, and click on **Expand**. This will expand all the nodes that were grouped into that metanode and show a convex hull around all the grouped nodes.

For example, the following screenshot shows the Les Misérables network when some of its nodes were grouped into different metanodes:

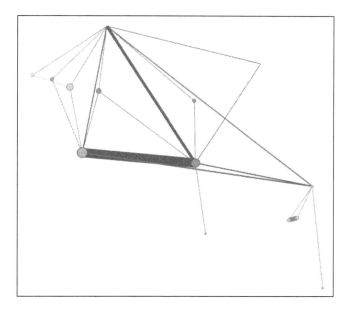

The following screenshot shows the same network when some of the metanodes have been expanded:

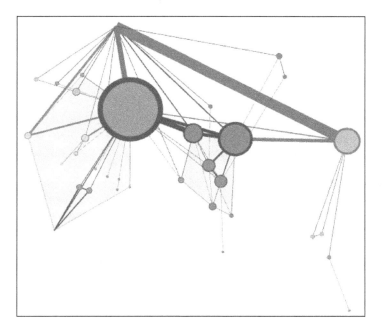

How it works...

A metanode is the node that is at the non-leaf level hierarchy in a multilevel tree. It replaces a bunch of nodes when they have been grouped together. Edges connecting metanodes with other nodes are called **metaedges**. These are derived based on the edges that were originally present between the nodes before grouping.

For example, consider the following screenshot:

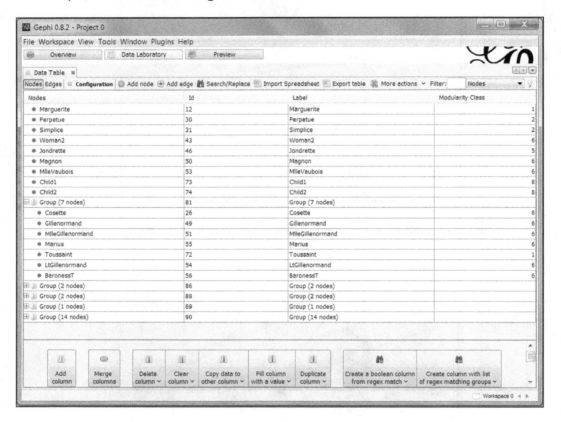

In the network for which the preceding screenshot was taken, the node with the **Group (7 nodes)** label is a metanode since it represents the group of nodes with the **Cosette**, **Gillenormand**, **MilleGillenormand**, **Marius**, **Toussaint**, **LtGillenormand**, and **BaronessT** labels.

Clustering links and attributes

One very useful feature of Gephi is the ability it provides to carry out clustering on links and attributes. This recipe will take you through this functionality in detail.

How to do it...

To understand how link- and attribute-based clustering happens in Gephi, follow these steps:

1. Load the Les Misérables network in Gephi.

2. Go to the **Data Laboratory** mode of Gephi by clicking on the **Data Laboratory** tab, which is located just below the menu bar.

3. In the **Data Laboratory** mode, in the upper-right corner of the Gephi window, adjacent to the **Filter** textbox, select **Modularity Class** from the drop-down list.

4. In the **Filter** textbox, enter 1, as shown in the following screenshot:

5. This will filter out all the nodes belonging to Modularity Class 1. Now, select all these nodes by clicking on the first one, holding down the *Shift* key and pressing the downward arrow button until all are selected, as shown in the following screenshot:

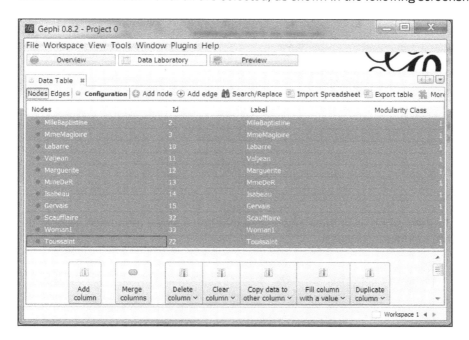

6. Now, as shown in the following screenshot, right-click and select **Group** from the drop-down list:

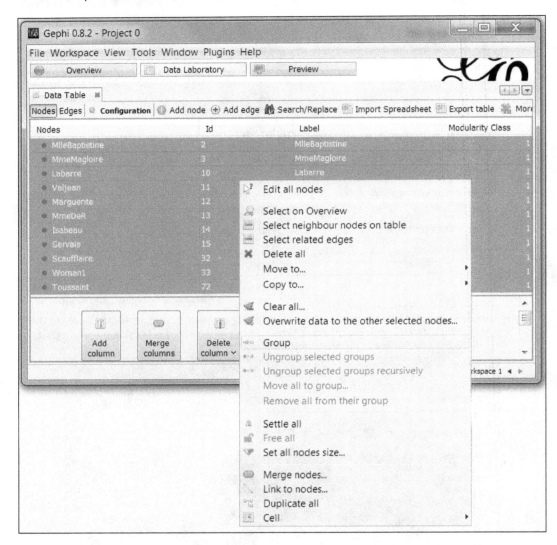

7. This will group all of these nodes together into a single metanode. Repeat this for all the modularity classes.

8. The following screenshot shows how **Data Laboratory** will look once this process is over. Click on the label of each metanode and allocate a custom name. Here, the nodes are named as *Modularity Class x* where *x* is the ID of the modularity class.

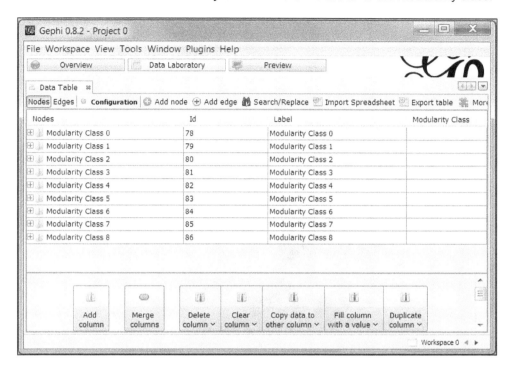

9. Go back to and run the Fruchterman Reingold Layout algorithm on this graph.

10. Next, go to the **Preview** mode and click on **Show Labels** under the **Node Labels** subtable.

11. Now uncheck the **Proportional size** checkbox and set an appropriate font for the node labels.

12. Click on the **Refresh** button to reload the new preview. The following screenshot shows how the graph will eventually look:

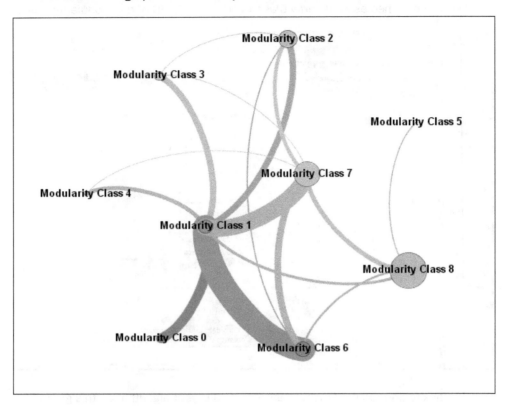

The same operations can be carried out on edges by clustering them according to certain attributes—for example, edge weight, edge target, and so on.

Getting Real-world Graph Datasets

In this chapter, we will cover the following recipes:

- ▸ Exploring the Web and Internet domain – EuroSiS Web mapping study
- ▸ Exploring the Web and Internet domain – the Internet dataset
- ▸ Exploring social networks – Zachary's karate club dataset
- ▸ Exploring social networks – Twitter's mentions and retweets dataset
- ▸ Exploring biological networks – the C. Elegans neural network dataset
- ▸ Exploring biological networks – the yeast dataset
- ▸ Exploring the infrastructure domain – the airlines dataset
- ▸ Importing data from MySQL databases
- ▸ Importing data from Neo4j databases
- ▸ Importing data via NodeXL

Introduction

Gephi has been used by researchers for exploring all sorts of networks. These networks range from social networks, biological networks, infrastructure domain networks, and so on. Apart from having domain-specific differences, these networks also differ in their size and structure. Social networks, for instance, are usually very large in size, whereas biological networks are considerably smaller, with different sorts of interactions. In this chapter, we are going to explore some of these networks in Gephi. We are also going to learn how to fetch data from different sources: MySQL databases, Neo4j databases, and NodeXL.

Exploring the Web and Internet domain – EuroSiS Web mapping study

In 2007, the EuroSiS mapping project was conceived. The project focused on mapping the interactions between Science in Society actors on the Web of 12 European countries. The mapping would thus reveal who are the most influential stakeholders in this network and who are the most active ones. In this recipe, we will use this network and explore it in Gephi.

Getting ready

To get started with this recipe, download the EuroSiS Web mapping study data from `https://gephi.org/datasets/eurosis.gexf.zip`. Once downloaded, unzip the downloaded file to obtain the EuroSiS Generale Pays file, which is in the GEXF graph file format.

How to do it...

The following steps describe how to analyze the EuroSiS data in Gephi:

1. Navigate to the folder where you unzipped the dataset and double-click on the `EuroSiS Generale Pays` file to load it in Gephi.

2. In the **Layout** panel, select the **ForceAtlas2** layout algorithm from the drop-down menu.

3. Select the **Dissuade Hubs** checkbox to push the hubs to the borders that lead to a clear demarcation between different classes of actors. The classes here refer to the actors from different countries. In other words, each border of a distinct class refers to the actors from the same country, and so on.

4. Select the **Prevent Overlap** checkbox to prevent overlaps in the graph. This just ensures that the final visualization that you see has different elements that are represented distinctly from each other for a better and clearer understanding and interpretation.

5. In the **Partition** panel, click on the **Nodes** tab and then on the **Refresh** button to reinitialize the list of partition parameters.

6. From the drop-down menu, select **country** as the partition parameter and hit **Apply**.

The following screenshot shows how the graph would look after these customizations:

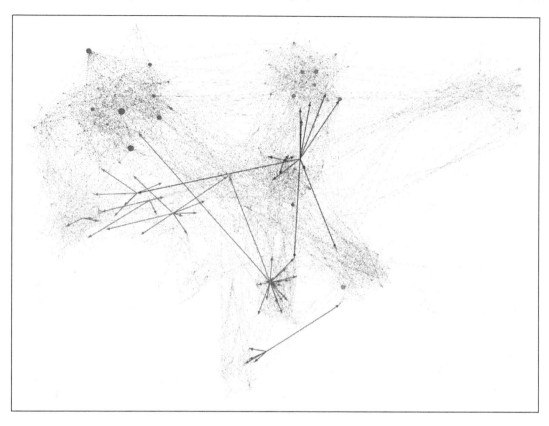

7. At the bottom of the panel, you will find the **Group** option. Click on that to group the nodes belonging to the same category into a single node.

8. Go to the **Preview** mode.

9. Tick the **Show Labels** checkbox in the **Node Labels** panel to show node labels.

10. Change the font of the node labels to make them look legible on the screen.

11. Change the node opacity from **100** to **1**.

12. Tick the **Box** checkbox to surround the node labels with a rectangular box the same color as the parent.

13. You can optionally choose to change the background of the preview for better visualization. Hit **Refresh** once you are done.

14. Zoom in into the preview for better clarity of the nodes and their labels.

   The following screenshot shows the graph after all these customizations:

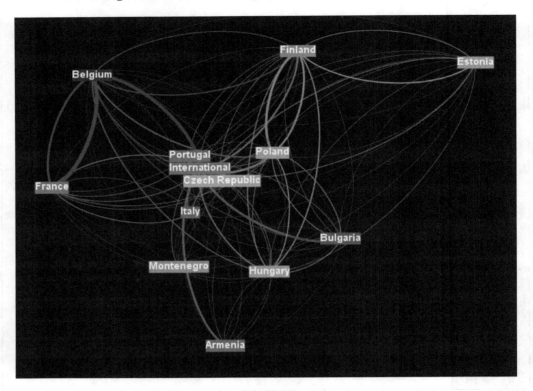

15. Navigate to the **Data Laboratory** mode. You will notice a series of tags that basically describe the domain(s) a particular actor is associated with.

16. To filter out actors in a particular domain, select the label of the column that holds information about that domain and type [1] in the **Filter** textbox. For example, if you chose **tag_energy**, this will only filter out actors in the energy domain.

17. You can also carry out the same filtering in the **Overview** mode. Go to the **Filter** panel.

18. Go to **Attributes** under **Library** and then go to **Equal**. Expand **Equal** and double-click on **tag_energy** to add this to the list of filter queries.

19. In the **Pattern** textbox towards the bottom of the screen, type in [1] and select the **Use Regex** checkbox. Hit **OK**.

20. Hit **Filter** to filter out the nodes that have **tag_enegry** equal to one.

How it works...

The EuroSiS mapping data is a brilliant dataset to start exploration with. It has some great node data with a wide variety of information. The node data consists of labels of nodes, the country information, the actor types, the association data with different domains, and so on. This is typically the dataset that you will come across in your day-to-day life—basically, a dataset with a lot of metainformation associated with it. This gives you a lot of opportunities for different kinds of analyses. The analysis is primarily driven by the type of information that you are looking out for. Here, we tried to view different associations and their strength among various countries present in this dataset by grouping organizations and other players based on the country they represent. This helped us look at this information from a very coarse, yet informative manner. We also looked at players in individual domains by filtering them out by using the metainformation present in this dataset in the form of tags. One can also look at various associations between actors in terms of their types. So, the bottom line is that the type of visualization that you lay out on this dataset and the elements that you choose to view depend on the problem statement that you are trying to solve and the cut of the dataset that you want to make in order to define boundaries for your exploration.

See also

▸ The final report of the EuroSiS mapping study from `http://www.eurosfaire. prd.fr/7pc/documents/1274371553_finalreporteurosis3_1.doc`

Exploring the Web and Internet domain – the Internet dataset

The case study that we are going to deal with in this recipe is that of the Internet dataset created in July, 2006 by Mark Newman. This data is a symmetrized snapshot of the structure of the Internet at the level of autonomous systems, reconstructed from BGP tables posted by the University of Oregon Route Views Project. An autonomous system is a network or a group of networks controlled by a common administrative entity. **Border Gateway Protocol** (**BGP**) is a standardized gateway protocol that is used for routing between autonomous systems on the Internet.

Getting ready

To get started with this recipe, download the required dataset from `https://gephi.org/datasets/internet_routers-22july06.gml.zip`. Unzip the dataset and load it in Gephi. You may want to make sure that you are using a powerful system to build this visualization. The Internet dataset is a pretty large dataset and requires a considerable amount of memory. Insufficient memory may lead to the system crashing.

How to do it...

To explore the Internet dataset that we have just downloaded, follow these steps:

1. Go to the folder where you extracted the dataset and double-click on the `internet_routers-22july06.GML` file to load it in Gephi.

   As the following screenshot of the context panel shows, this is a pretty big network, consisting of nearly 23 thousand nodes and nearly 49 thousand edges between those nodes:

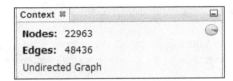

   The network is not comprehensible to begin with, due to the large amount of data present. To make it comprehensible, we will partition it based on degree information.

2. In the **Statistics** panel, hit **Run** against the **Average Degree** metric.

3. Once done, go to the **Partition** panel and hit the **Refresh** button to populate the drop-down menu with a list of partition parameters.

4. Select **Degree** from the drop-down list.

5. Hit **Apply** to apply the colors to various partitions.

6. In the Layout panel, select the **ForceAtlas2** layout algorithm with the **Dissuade Hubs** and **Prevent Overlap** options on in order to get a better visualization of the graph.

The following screenshot shows the Internet graph in the **Preview** mode after the previous customizations:

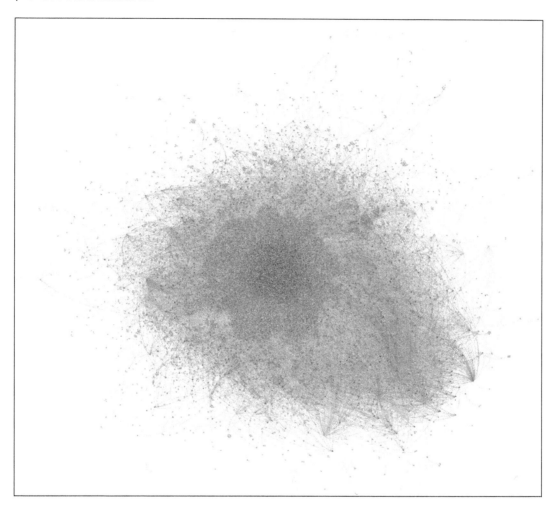

How it works...

The Route Views project was conceived to act as a tool to understand the global routing system. The Internet dataset, as mentioned in the introduction, is a snapshot of the data collected as part of this system. As mentioned before, this is a pretty huge dataset with around 23 thousand nodes and 49 thousand edges between those nodes. As can be seen from the final visualization that we obtained, there is a core of routers that are highly connected with others in the system and sit in the center of this system. They have very high degrees and are also closely connected to other high-degree routers in the system. Then there are routers with degrees in the medium range and they are mostly connected to the high-degree or central routers in the system. They also act as bridge from the sparsely connected routers in the system to the core of this network. Finally, there are a lot of routers that are not very well connected and sit on the periphery of the system. They are connected via bridges or intermediate nodes, in other words, to the rest of the network.

There's more...

In the *Finding the HITS value for a graph* recipe in *Chapter 5*, *Running Metrics, Filters, and Timelines*, the concept of hubs and authorities was introduced. Hubs and authorities basically help in understanding how central and how informative various nodes in a network are. You can run the HITS algorithm from the **Statistics** panel to generate the hubs and authorities score for the nodes in this system. You can then rank the nodes based on either of these scores and color them from the **Ranking** panel in the **Overview** mode. You can, additionally, look at filtering out the nodes with a certain hub/authority score.

See also

▶ http://www.routeviews.org/ to know more about the University of Oregon Route Views Project

Exploring social networks – Zachary's karate club dataset

In 1977, Wayne W. Zachary published a paper in the *Journal of Anthropological Research* titled *An Information Flow Model for Conflict and Fission in Small Groups*. In this paper, he wrote about a social network of friendships between 34 members of a karate club at a US university. This paper depicted the information flow model for conflict and fission in small groups. In this recipe, we are going to dig into this case study and get a better understanding of this dataset.

Getting ready

To get started with this recipe, download the Zachary's karate club dataset from `https://gephi.org/datasets/karate.gml.zip`. Unzip the downloaded ZIP file and load the `gml` file into Gephi.

How to do it...

Follow these steps to explore the Zachary's karate club dataset that you just downloaded and unzipped:

1. Navigate to the folder where you unzipped the dataset and double-click on the `karate.GML` file to load it into Gephi.

2. You will notice that this is a very small network with just 34 nodes and 78 edges. To begin with, the nodes haven't been colored and, hence, it is a bit difficult to make a much sense of this data. Run the **Average Degree** metrics from the **Statistics** panel.

3. Go to the **Ranking** panel and, from the drop-down menu, select **Degree** in order to rank the nodes according to their degrees.

4. Choose an appropriate color palette and hit **Apply** once you are done.

5. Go to the **Preview** mode and check the **Show Labels** checkbox under the **Node Labels** panel.

6. Set an appropriate font for the node labels.

7. Tick the **Box** checkbox to show a rectangular box around the node labels.

8. You can optionally choose to set a background color for the graph preview.

9. Hit **Refresh** to generate the preview once you are done. The following screenshot shows the final graph in the **Preview** mode. The purple nodes are the ones with the highest degree whereas the dark orange nodes are the ones with the lowest degree. The rest of the nodes have degrees in the intermediate range.

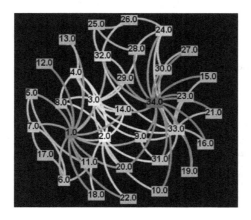

10. Go back to the **Overview** mode and hit **Run** against **Eigenvector Centrality** in the **Node Overview** tab under the **Statistics** panel to find the eigenvector centrality score for the nodes in this vector.

11. Go to the **Ranking** panel and choose **Eigenvector Centrality** as the ranking parameter under the **Nodes** tab.

12. Choose an appropriate palette and hit **Run**.

13. Go to the **Preview** mode and repeat steps 5 to 9 in order to generate a preview for the new visualization of the graph.

The following screenshot shows the new graph:

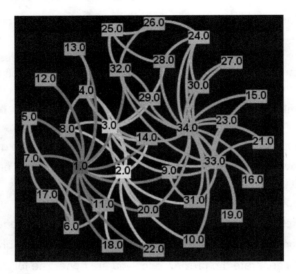

How it works...

As already mentioned in the introduction, the Zachary club dataset depicts a social network of friendships, conflicts, and fissions between 34 members of a karate club at a US university. This is clear from the two graphs that you have seen in this recipe. There are certain members who are quite popular and favored in this group, whereas there are others who have just one or two connections with the rest of the group.

Something that is interesting to study is how two main yet not disjoint groups have formed and correspond to the folks represented by the node labeled **1.0** and the one labeled **34.0**. This is clear from the first graph in this recipe. These two have the maximum number of connections, a lot of them being mutual. This kind of behavior is evident in a lot of real-world networks and is usually referred to as the process of **community detection**.

Another very interesting fact that can be noticed when comparing the first and the second graph in this recipe is that the one with the maximum number of connections is not always central to the network. For example, the node with label 34.0 has one of the highest connections in this network but doesn't have a high eigenvector centrality score. On the other hand, the node with label 1.0 has very high number of connections, as well as a high eigenvector centrality score. This means that the node with label 1.0 not only has a lot of connections but a lot of friendships in this network have this node as their intermediate node. This means that, for someone who is studying the importance of people in this network, node 1.0 will be of much more interest than node 34.0. In some texts, node 1.0 is referenced as the connector node.

There's more...

The concept of connectors that was introduced towards the end of the *How it Works...* section was introduced by a social scientist and journalist Malcolm Gladwell in his book titled *The Tipping Point: How Little Things Can Make a Big Difference*. He talks about different sorts of people who exist around us in the society and how some of them play crucial roles in dissemination of the information. He categorizes these people into three categories: Connectors, Mavens, and Salesmen. The book is available for purchase from `http://www.amazon.com/The-Tipping-Point-Little-Difference/dp/0316346624`.

See also

▸ The paper titled *An Information Flow Model for Conflict and Fission in Small Groups* by Wayne W. Zachary, published in the *Journal of Anthropological Research* in 1977, to learn more about the dataset covered in this recipe. The paper can be downloaded from `http://www1.ind.ku.dk/complexLearning/zachary1977.pdf`.

▸ The report titled *Community detection in graphs* by Santo Fortunato, published in Physics Report in 2010, at `http://arxiv.org/abs/0906.0612`.

▸ The paper titled *Community detection in large-scale social networks* by Nan Du, Bin Wu, Xin Pei, Bai Wang, and Liutong Xu, published in the Proceedings of the ninth WebKDD and first SNA-KDD workshop on Web mining and social network analysis in 2007, at `http://dl.acm.org/citation.cfm?id=1348552`.

Exploring social networks – Twitter's mentions and retweets dataset

Twitter is one social network that is explored the most by researchers. The sort of data that is available on Twitter makes it a lucrative source for social network analysis. In this recipe, we are going to consider a small dataset of Twitter's mentions and retweet and explore it in Gephi.

Getting ready

Download Twitter's mentions and retweet dataset from the URL `http://rankinfo.pkqs. net/twittercrawl.dot.gz`. Unzip the dataset to start exploring it. You may want to make sure that you use a powerful system to build this visualization. This dataset is a pretty large dataset and requires a considerable amount of memory. Insufficient memory may lead to the system crashing.

How to do it...

To analyze the Twitter mentions and retweets data that you just downloaded, follow these steps:

1. Start Gephi.

2. From the **File** menu, click on **Open** and navigate to the folder where you extracted the Twitter dataset. Open the `twittercrawl` file in Gephi.

3. In the **Statistics** panel, click on **Run** against **Average Degree** to generate the degree information for each of the nodes in the network.

4. Go to the **Partition** panel and click on the **Refresh** button to populate the drop-down menu with a list of partition parameters.

5. From the drop-down menu, select degree as the partition parameter and hit **Apply** to apply the colors to the graph.

6. From the **Layout** panel, run the **ForceAtlas2** layout algorithm with the **Dissuade Hubs** checkbox and the **LinLog** mode on for better visualization.

7. You may need to increase the scaling in order to make sure the final graph doesn't shrink down a lot. To do so, click on the textbox next to **Scaling** and enter a big value in there, something like 30. Hit **Run** to run the layout algorithm.

The following screenshot shows how the graph will look in the **Preview** mode at the end of the algorithm execution:

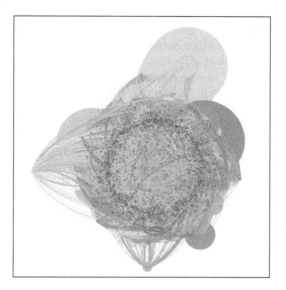

8. You can also filter out the nodes with the highest out-degree by selecting **Out degree** from the **Range** option under **Attributes** in the **Filter** panel.

The following screenshot shows the Twitter network in the **Preview** mode when the nodes with out-degree in the range 360 to 1051 have been filtered out:

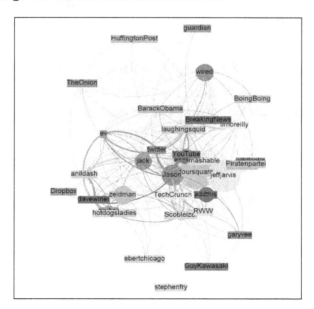

How it works...

The Twitter retweets and mentions dataset is an example of a large-scale network. At first sight, it might resemble the Internet network dataset that we analyzed earlier in this chapter, but it is actually a lot different from that dataset. If you notice, there are fewer nodes, around 3,000, but the number of edges between these nodes is much larger than in the Internet dataset. There are around 188,712 edges present in this dataset. The nodes in this dataset refer to different users or accounts on Twitter and the directed edges represent the fact that the source user has either mentioned the destination user in his/her tweet or he/she has retweeted a tweet by the destination user.

A lot of nodes in this network have large in-degrees and out-degrees. There is a single component in this network, which means that every entity in this network has a relationship with at least one other entity and none of these are disconnected. Sorting the data with respect to the degree column would reveal that accounts such as Foursquare, Twitter, Techcrunch, Techmeme, and so on are the most mentioned accounts in Twitter. This isn't a surprise since these accounts have a large number of followers who either mention these accounts quite often or retweet the tweets from these accounts. Another thing to notice is that even though these accounts have a high degree, they have high in-degree but relatively low out-degree. This means that they do not quote others very often and act more like information disseminators in the social network.

There's more...

There has been a plethora of research on the social network, Twitter. Researchers have studied information dissemination, community formations, sentiment analysis, political alignments, and so on. Take a look at `scholar.google.com` and search for Twitter to look at a long list of research articles, papers, and reports published in this regard.

Exploring biological networks – the C. Elegans neural network dataset

C. Elegans, which is short for **Caenorhabditis Elegans**, is a free-living roundworm that lives in soil environments. Since 1960s, this organism has been widely studied by scientists because of the fact that it is the simplest organism with a nervous system. In this recipe, we will explore the neuron network of C. Elegans and study one of the most common types of networks called the small-world network; these are very often found in the world around us.

Getting ready

To get started with this recipe, download and unzip the C. Elegans neural network data from `http://gephi.org/datasets/celegans.gexf.zip`.

How to do it...

To explore the C. Elegans neural network data and investigate its interesting properties, follow these steps:

1. Open Gephi. Click on **File** in the menu bar and then **Open** in the drop-down menu.

2. In the **Open** dialog window, navigate to the folder where you unzipped the C. Elegans dataset and click on the `celegans.gexf` file. Hit **Open** once done.

3. In the **Import** report, choose **Directed** in the **Graph Type** drop-down menu and hit **OK**.

4. In the **Layout** window, choose **ForceAtlas2** from the **Layout** drop-down menu.

5. In the **Threads** option, set the number of process threads according to your processor's capacity.

6. Under **Behavior Alternatives**, check the checkboxes next to **Dissuade Hubs**, **LingLog mode**, and **Prevent Overlap**.

7. Under **Tuning**, tick the **Stronger Gravity** checkbox.

8. Hit **Run** once done. After the algorithm's execution is over, the C. Elegans network will look similar to the one shown in the following screenshot:

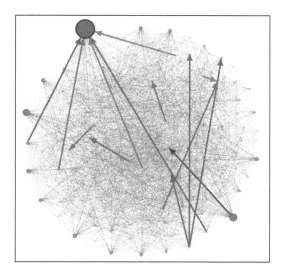

9. In the **Statistics** panel on the right side of the Gephi application window, click on **Run** next to **Avg. Path Length**. Once the execution is over, you will see that the result is **3.992**. This means that, on an average, about four hops are required to traverse from one node to the other in the network.

10. In the **Statistics** panel, click on **Run**, which is next to **Average Degree**. The returned result will be **7.663**. This means that, on an average in this network, each node has about seven to eight neighbors.

11. Under **Node Overview** in the **Statistics** panel, click on **Run** next to **Avg. Clustering Coefficient**. This will return **0.164** as the result.

How it works...

Small-world networks were first described by Duncan J. Watts and Steven H. Strogatz in their paper titled *Collective Dynamics of Small-World Networks* that was published in a journal titled Nature in 1998. According to Wikipedia, a small-world network is a type of network in which most of the nodes are not neighbors of one another, but are reachable from every other by a small number of hops. In other words, there is a small distance to be covered while traversing from any node to any other node, even though they aren't immediate neighbors of each other. This distance is a function of the number of nodes in the network. To be precise, the typical distance between any two nodes in a small-world network is directly proportional to the logarithm of the number of nodes in the network.

C. Elegans' neural system comprises around 300 neurons. A comprehensive wiring diagram of these neurons is called a **connectome** and it has been found to be small-world network in C. Elegans. This means that even though most of the neurons in C. Elegans are not connected to each other, each of them is reachable from any other by a very small number of hops. This is evident by the result we got by running the average path length metric on the C. Elegans network in Gephi. The average path length came out to be 3.992, which is fairly small for a network with 306 nodes and 2,345 edges. Another property that we just now learned is the existence of small number of immediate neighbors. We saw the nodes in this network have on average about seven to eight neighbors. In fact, on applying the degree filter from the **Filters** panel, you will find that around 65 percent of the nodes have degrees in the range 0-15.

Yet another property that a small-world network has is a significantly higher average local clustering coefficient in comparison to a random graph built on the same vertex set. The local clustering coefficient for a node in a graph is defined as the ratio of the links that exist between the node and its neighbors to the total number of links that are possible between this node and its neighbors.

The local clustering coefficient for a vertex v_i in a directed graph is given as follows:

$$C_i = \frac{\left|\left\{e_{jk} : v_j, v_k \in N_i, e_{jk} \in E\right\}\right|}{k_i\left(k_i - 1\right)}$$

Here, e_{jk} is the edge between vertex v_j and v_k. N_i is the neighborhood of vertex v_i and k_i is the number of vertices in N_i. The average local clustering coefficient for the graph is the average of the local clustering coefficients of each of the vertices in the graph.

We saw that the average local clustering coefficient for the C. Elegans network came out to be 0.164. To get an estimate of the average local clustering coefficient of a random graph with a similar vertex set, generate a random graph in Gephi by clicking on the **File** option in the menu bar followed by **Generate** and then **Random Graph**. In the pop-up window, enter the number of nodes as `306` and the wiring probability as `0.0495`. This will generate a random graph with edges somewhere equal to the number of edges in the C. Elegans network. You might have to repeat this process multiple times to get the desired graph. On running the **Avg. Clustering Coefficient** metric for this random graph, the result will be somewhere around **0.024**, which is significantly smaller than the average local clustering coefficient of the C. Elegans neural network.

See also

▸   The paper titled *Collective Dynamics of Small-world Networks* by Duncan J. Watts and Steven H. Strogatz, which was published in the *Nature* journal in 1998, to know more about the small-world networks

▸   `http://www.scholarpedia.org/article/Small-world_network` to know more about small-world networks

▸   *Chapter 20, The Small World Phenomenon*, in the book titled *Networks, Crowds and Markets: Reasoning about a Highly Connected World* by David Easley and Jon Kleinberg at `http://www.cs.cornell.edu/home/kleinber/networks-book/networks-book-ch20.pdf`

Exploring biological networks – the yeast dataset

In 2003, Shiwei Shun, Lunjiang Ling, Nan Zhang, Guojie Li, and Runsheng Chen published a paper titled *Topological structure analysis of the protein-protein interaction network in budding yeast* in *Nucleic Acids Research*. As the title of the paper very well conveys, the study involved an analysis of the different interactions between various proteins that are present in yeast. In this recipe, we will consider this protein-protein interaction network of yeast and study its properties. So, let's dive in.

Getting ready

Download the network dataset from `https://gephi.org/datasets/yeast.gexf.zip` and unzip it into a folder.

How to do it...

The following steps illustrate how to explore and study the properties of the protein-protein interaction network of yeast.

1. Navigate to the folder where you unzipped the downloaded yeast network. You will find a file named `yeast.gexf`.

2. Double click on the `yeast.gexf` file to open it in Gephi. This is how the network will look once it is loaded in Gephi:

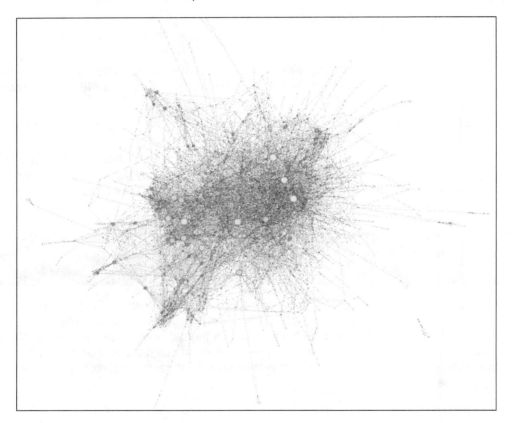

3. To get a better and more legible view of the network, go to the **Layout** panel, select **ForceAtlas2** from the drop-down menu, select **Dissuade Hubs** and **Prevent Overlap**. Set **Scaling** to **200** and **Gravity** to **2000**. Hit **Run**.

The following screenshot shows the yeast network after the ForceAtlas2 layout algorithm has been run on it with the prescribed settings:

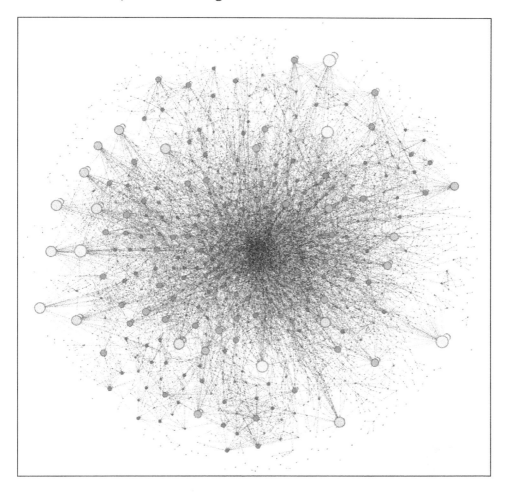

1. You may have noticed a number of scattered nodes around the periphery that indicate the presence of disconnected components in the network. Click on **Run** against **Connected Components** in the **Statistics** panel.

2. One interesting property to check will be the small-world property. To do so, run the **Average Degree** and **Average Path Length** metrics from the **Statistics** panel.

How it works...

The result obtained upon running the Connected Components statistic on the yeast's protein-protein network the result obtained would be 59. This means that there are quite a lot of disconnected components in this network. In other words, this shows the presence of very tiny protein groups that interact with themselves and not with the other proteins in this network. To confirm this, in the **Partition** panel, go to **Nodes** and click on the **Refresh** button. This will populate the drop-down menu with a list of partitioning parameters. One of these will be **Component ID**. Choosing **Component ID** from the drop-down menu will show a list of components present in the network, along with their IDs and the share of the whole network they represent. You will notice that the component with ID 58 covers about 94.2 percent of the entire network and the remaining 5.8 percent is covered by the other 58 components. This means that 94.2 percent of the total proteins interact with each other in some way or another, but the remaining 5.8 percent have their interactions in isolation.

Running the average path length, average degree, and average clustering coefficients would give 4.648, 3.042, and 0.065 as results respectively. A small path length, a small degree, and a significantly high clustering coefficient, to some extent, point to the presence of a small-world property in this interaction network.

See also

▶ The paper titled *Topological structure analysis of the protein-protein interaction network in budding yeast* by Shiwei Shun, Lunjiang Ling, Nan Zhang, Guojie Li, and Runsheng Chen, published in *Nucleic Acids Research* in 2003, to know more about the analysis of yeast's protein-protein interaction network. The paper can be found at `http://library.ibp.ac.cn/html/slwj/000182627700029.pdf`.

Exploring the infrastructure domain – the airlines dataset

So far, we have covered various domains, including biology, social networks, and the Internet, for our network analysis. In this recipe, we are going to look into yet another interesting domain—transportation networks. For the purposes of this recipe, we are going to consider the airlines data from the North American Transportation Network Data that was published in 2008.

Getting ready

To get started, download the dataset from `https://gephi.org/datasets/us-air97.net.zip` and unzip it into a folder.

How to do it...

The following steps illustrate how to analyze the dataset that we just downloaded:

1. Navigate to the folder where you unzipped the dataset and double-click on the USAir97.NET file to load it into Gephi.

2. Under the **Layout** panel, choose the **ForceAtlas2** layout algorithm.

3. Tick the **Dissuade Hubs** and **Prevent Overlap** checkboxes. Hit **Run**.

4. Under the **Ranking** panel, choose **Nodes** and then **Degree** from the drop-down list. Choose a color palette and hit **Apply**.

   The following screenshot shows how the graph looks after step 4:

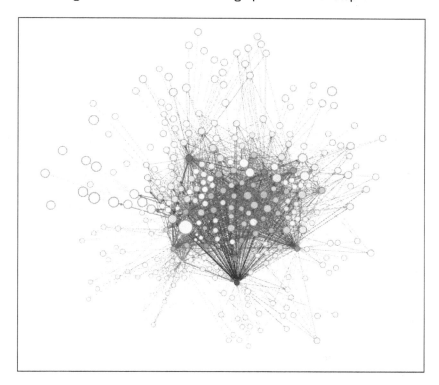

5. Under the **Network Overview** panel, click on **Run** against **Average Degree**.

6. Go to the **Data Laboratory** mode and then click on **Nodes** to reveal the nodes data. You will notice the extra column titled **Degree** there. Here's how the data will look:

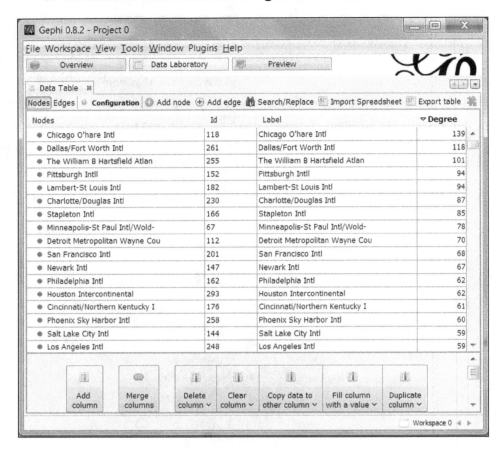

7. Now go back to the **Overview** mode and navigate to the **Filters** panel.

8. In the **Filters** panel, go to **Attributes**, followed by **Range** and then **Degree**. Double-click on **Degree** to add it to the filter queries list.

9. Move the **Range (Degree) Settings** slider to choose a range of airports. Let us say, for instance, the range chosen is from 33 to 139.

10. Click on **Filter** to filter out nodes that are outside of this range and their corresponding edges.

11. Expand the bottom panel and check the **Node** checkbox under the **Label** tab to show node labels.

12. Run the Fruchterman Reingold algorithm on this graph, followed by the Label Adjust algorithm.

13. Go to the **Preview** mode.

14. Tick the **Show Labels** checkbox under the **Node Labels** panel and hit **Refresh**. You will see a graph similar to the one shown in the following screenshot:

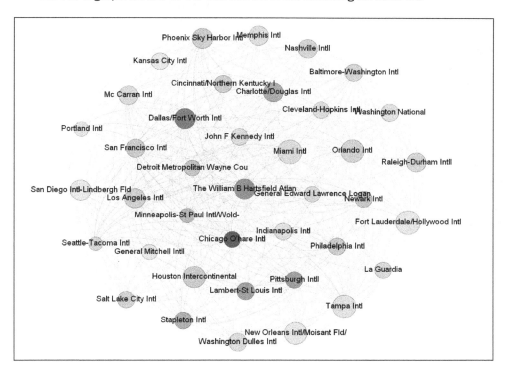

How it works...

The airport dataset from North American Transportation Atlas Data is a point database of public-use landing facilities in the 50 states and U.S. territories in the United States in 1997. If you take a proper look at the first graph in this recipe, you will notice that some of the airports sit right at the center of this dataset with a huge number of interlinks with other airports. These are the hubs of this network. Some airports surround these airports with a lesser number of interlinks. The rest of the airports, placed around the periphery and marked in cream, have very few links with the other airports. This is perfectly aligned with how one would visualize an airport network to be. There are some very busy airports that have flights operating to most of the other airports in the network, some airports operate at normal frequencies, and then there are some that operate at very low frequencies and have very few flights to a couple of the other airports. A look at the node data sorted by degree in the **Data Laboratory** mode would reveal that Chicago O'hare International Airport, Dallas/Fort Worth International Airport, and The William B. Hartsfield Atlanta Airport are the top three busiest airports. On the other hand, there are airports such as Napakiak, Eek, Gustavus, and so on, that only operate flights to a single airport.

Importing data from MySQL databases

So far, we have learned how to work with graph or network data present in the form of **comma-separated value** (**csv**) files, GEXF files, GML files, and so on. These are all ways of explicitly gathering data into files that are directly ingestible by Gephi. One really cool feature of Gephi is its ability to ingest data directly from sources such as MySQL databases. In this recipe, we will learn how to get network/graph data from the MySQL database and consume it directly in Gephi for analysis.

Getting ready

To perform this recipe, ensure that you have a MySQL database set up on your system and running. The installation of MySQL databases is beyond the scope of this book but there is enough documentation on the Internet to help you with that. Apart from that, ensure that you have a database created for your use.

How to do it...

To learn how to directly consume network data from MySQL into Gephi, follow these steps:

1. Start Gephi and navigate to **File** in the menu bar.

2. Click on **Import Database** in the drop-down list.

3. Click on **Edge List** from the pop-up option. A new window, as shown in the following screenshot, will appear:

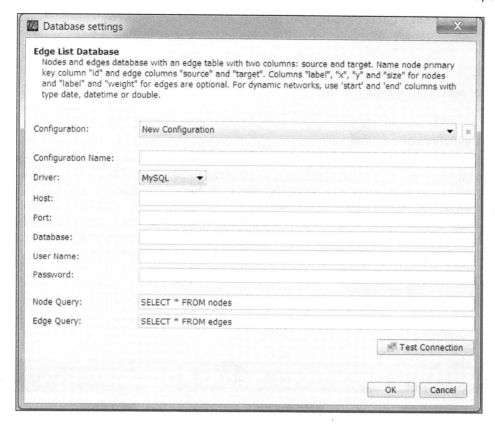

4. In the **Configuration Name** field, enter the name that you want to give to this configuration.

5. Select the type of driver from the **Driver** drop-down menu.

6. Enter the host IP address and port number in the **Host** textbox and **Port** textbox respectively.

7. Enter the name of the database in the **Database** textbox.

8. Enter your authorization credentials (your username and password) in to the appropriate boxes.

9. In the **Node Query** textbox, enter the SQL query that will fetch the node information from the database. Keep in mind the instructions that have been given towards the top of this window.

10. In the **Edge Query** textbox, enter the SQL query that will fetch the edge dataset from the database. As mentioned in the previous point, keep in mind the instructions given towards the top of this window.

11. You can optionally choose **Test Connection** to ensure that everything works fine.

12. Hit **OK** once you are done.

How it works...

MySQL is one of the most widely used relational database management systems (**RDBMS**) in the world. The information or the data is represented as tables. Each row in a table represents a data object with different properties or information. If this information consists of entities and relationships between them or, in other words, if the information contains some form of network data, it can be easily consumed by Gephi. The entity information goes in as nodes and the relationship information gets modelled as edges. The process is pretty straightforward, thereby adding another star to the great list of functionalities offered by Gephi.

See also

▶ http://www.tutorialspoint.com/mysql/ for a great series of tutorials on MySQL

Importing data from Neo4j databases

There's a new generation of databases called **graph databases** in which all the entities are represented in graphical forms, consisting of nodes and edges. One such database is Neo4j. Gephi has the ability to consume data from the Neo4j database. In this recipe, you are going to learn about this.

Getting ready

Download and install Neo4j on your system. Also, download a sample dataset from http://neo4j.com/developer/example-data/. This page also has instructions to load this dataset into Neo4j.

How to do it...

The following steps illustrate how Gephi consumes data from Neo4j:

1. Start Gephi and click on the **Plugins** option under the **Tools** option in the menu bar. This opens up a **Plugins** window, as shown in the following screenshot:

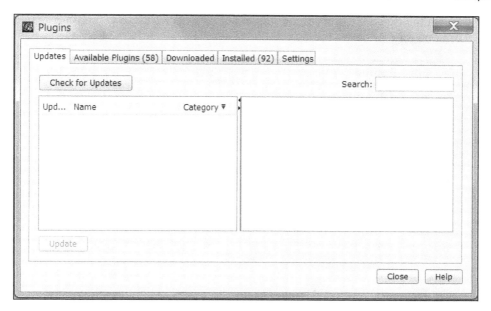

2. Click on the **Available Plugins** tab to view a list of available plugins for download.

3. Tick the checkbox that is located right next to **Neo4j Graph Database**.

4. Hit **Install**. This installs the Neo4j database plugin in Gephi.

5. Now click on the **File** option in the menu bar and select **Neo4j Database**.

6. Select **Full Import** from the list. This opens up a **Choose source Neo4j database** window, as shown in the following screenshot:

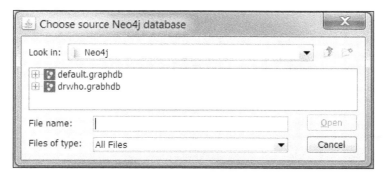

7. Select the database that you want to import and click on **Open**. The **Choose traversal options** window opens up:

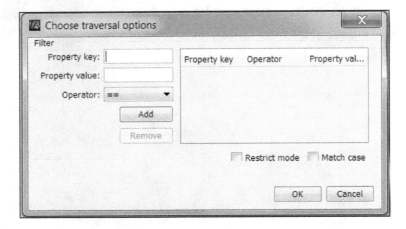

8. Enter the property key and property value for import in the respective fields. Choose the required operator. If you do not want to do any filtering on this dataset, just leave these fields blank and hit **OK** to load the data into Gephi.

How it works...

Neo4j is one of the most popular graph databases in the industry. It is implemented in Java and is licensed under the free GNU General Public License v3. Neo4j uses a declarative query language known as Cypher Query Language for expressive and efficient querying and updating of the graph store.

Being a graph database, Neo4j stores everything in the form of nodes, edges, and attributes. Each node and edge can have any number of attributes. You can also have labeled nodes and edges. Gephi's Neo4j Graph Database plugin understands this format and hence directly translates the data from Neo4j into a format that Gephi can understand understandable by Gephi.

See also

- ▶ http://neo4j.com/developer/get-started/ to learn more about Neo4j
- ▶ *Learning Neo4j* by Rik Van Bruggen at http://neo4j.com/book-learning-neo4j/
- ▶ *Graph Databases* by Ian Robinson, Jim Webber, and Emil Eifrém, O'Reily Media at http://neo4j.com/graph-databases-a/

Importing data via NodeXL

Yet another popular source of data import apart from Excel, Neo4j, and MySQL databases is NodeXL. NodeXL is a free, open source template for Microsoft Excel that lets its user work with graphs. In this chapter, we will learn how to import data into Gephi via NodeXL.

Getting ready

The explanation of how to prepare data in NodeXL is beyond the scope of this book. You can refer to `http://social-dynamics.org/twitter-network-data/` for details on this. Using the same instructions, I have prepared a dataset of tweet data from Twitter. The tweets are picked up from the public search stream related to the word Gephi.

The following screenshot shows the vertices data in NodeXL:

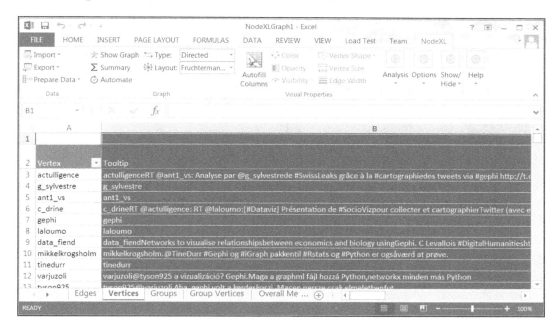

The following screenshot shows the edges data in NodeXL:

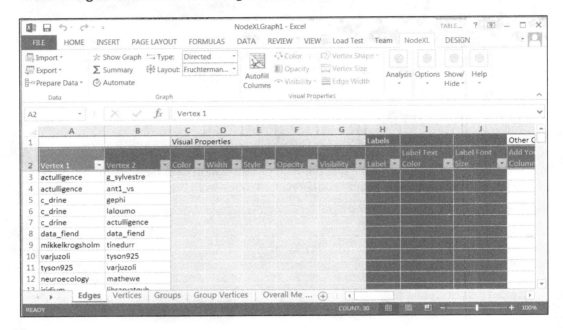

How to do it...

In order to import data from NodeXL into Gephi, follow these steps:

1. Save the data from NodeXL into a GraphML file by clicking on **NodeXL**, followed by **Export**, which is located in the top-left corner, and then **To GraphML file**.

2. Load this file into Gephi. This is how the node data will look in the Gephi's **Overview** mode:

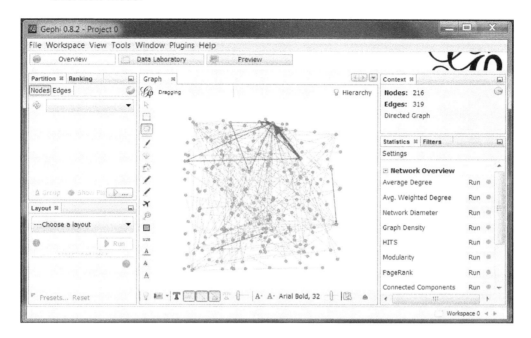

3. You can click on the **Data Laboratory** mode and view details about the data there.

How it works...

NodeXL ingests and analyzes graphical data that is represented in terms of vertices and edges separately. When one saves this data in GraphML format, it becomes directly consumable by Gephi. Gephi loads the vertices data as node data and the edge data as its own edge data, and constructs a graph.

See also

▸ The book titled *Analyzing Social Media Networks with NodeXL* by Derek L. Hansen, Ben Shneiderman, and Marc A. Smith to learn about NodeXL. The book is available at `http://www.sciencedirect.com/science/book/9780123822291`.

10
Exploring Some Useful Gephi Plugins

In this chapter, we will cover the following recipes:

- ▸ Exporting networks on the Web using Seadragon
- ▸ Exporting rich interactive visualizations using Sigma.js
- ▸ Describing complex network structures using GEXF
- ▸ Generating world maps
- ▸ Performing social network analysis

Introduction

Starting from a tour of Gephi and its features, various layout algorithms, statistical metrics, filters, and the Data Laboratory mode, and moving on to getting our hands dirty working with real-world datasets, we have covered almost everything that Gephi has to offer. We gained a good understanding of how to create and manipulate a new graphical network, as well as exploring and manipulating an existing network. But sometimes all of this is not sufficient for everything that we want to accomplish in our data exploration. In fact, it is not even feasible for a tool to have all the features one could think of. That is where plugins come into the picture. Gephi has a really helpful marketplace where users can find a lot of useful plugins that can be used alongside Gephi to carry out whatever the user needs to on a dataset. In this chapter, some of the plugins that are extensively used by researchers and developers working with Gephi are covered. So, let's jump right in.

Exporting networks on the Web using Seadragon

In the previous chapters, we learned how to export a network or graph built in Gephi to a PDF, SVG, or PNG file. Sometimes, this type of export might not be useful. For instance, if one were to display the graph on a Web page, a better rendering mechanism than the static rendering offered by the PDF, SVG, or PNG format is required. Gephi Marketplace has a really good plugin to solve this problem. Seadragon is a plugin available on Gephi Marketplace that allows networks built in Gephi to be directly exported and used on a browser. Networks are exported as image tiles and this makes it very easy and convenient for someone to study them in detail. In this recipe, we will learn how to install and use this plugin.

Getting ready

In order to export a graph built in Gephi to a format that is directly consumable by a browser, the Seadragon plugin has to be installed first. The following steps illustrate how to install this plugin:

1. Open Gephi.

2. Go to **Tools | Plugins**. This opens up the **Plugins** window, as shown in the following screenshot:

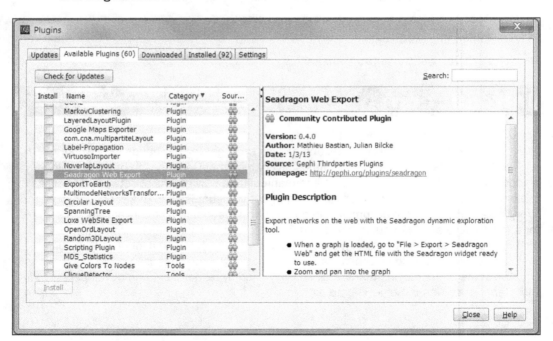

3. Click on the **Available Plugins** tab and search for **Seadragon Web Export** in the list of available plugins.

4. Check the checkbox next to **Seadragon Web Export** and hit **Install**.

5. Once the installation is complete, a new window asking to restart Gephi opens up. Click on **Restart Now** and then **Finish**. This installs the Seadragon plugin for Gephi.

How to do it...

In the following steps, the process of using the Seadragon Web Export plugin for exporting networks built in Gephi is explained:

1. Start Gephi and, from the **Welcome** screen, load the Les Misérables network.

2. Navigate to the **Preview** mode.

3. Select the checkbox next to **Show Labels**, which is under the **Node Labels** panel.

4. Click on **File** in the menu bar.

5. Navigate to **Export | Seadragon Web**. This opens up the **Seadragon Web Export** window, as shown in the following screenshot:

6. Enter the path of the folder where you want to export the network.

7. Enter the width, height, tile size, and margins for the export. Note that a large value for width and height will lead to high-quality graphics but might result in a very long export time; the application might even hang due to memory limitations.

8. Hit **OK**. Once the export is over, the following message box opens up:

9. Click on **Open in browser** to load the exported image into a browser. The following screenshot shows the Les Misérables network exported to a web browser:

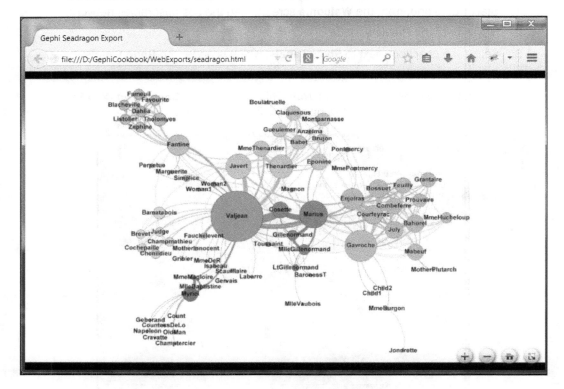

10. You can click on the **+** and **-** signs located at the bottom-right corner of the browser window to zoom in and zoom out the graph.

How it works...

Seadragon is a Web-optimized technology that allows seamless browsing of graphics and photos on a browser. It is a Microsoft Live Labs application and powers many different platforms. Seadragon technology makes is possible to smoothly view extremely large and high-resolution images with almost negligible latency. Seadragon for Gephi is a plugin that allows users to export very high-resolution and large graphs to the Web for viewing on browsers.

Seadragon is implemented in pure JavaScript and works on all the modern browsers. It uses image tiles, which basically means that the image is subdivided into a regular grid of smaller images in order to exploit the local spatial coherence in the image. This comes in handy when there are hardware limitations and the entire high-resolution image cannot be rendered completely at once. This process is also known as **Tiled Rendering**. Since the graph is divided into image tiles, each of which is rendered one at a time, there's no limit on the size of a graph in Gephi. One thing to notice here is that all the enhancements made in the **Preview** mode are preserved when the network is exported using Seadragon.

There's more...

If you are using Chrome as the browser, you will notice an error being thrown up when you try to load the exported network. This is due to a bug that exists in Chrome that prevents files the on a user's machine from being loaded into the browser. The workaround to this problem is to go to **Windows Command**, navigate to the path in which Chrome is installed and then run Chrome with the **—allow-file-access-from-files** flag option. The details can be found at `http://chrome-allow-file-access-from-file.com/`.

See also

▸ The Wikipedia page for Seadragon at `http://en.wikipedia.org/wiki/Seadragon_Software` for more detail on this technology

▸ `https://marketplace.gephi.org/plugin/seadragon-Web-export/` to know more about the Seadragon plugin for Gephi

Exporting rich interactive visualizations using Sigma.js

In the previous recipe, you learned how to export networks built in Gephi to a Web browser by using the Seadragon Web Export plugin, which is available from Gephi Marketplace. The network thus exported allows users to look at the high-resolution images tile-by-tile and explore the network in detail. But one limitation that exists there is the absence of interactivity with the graph. The user cannot interact with the graph and study its properties; rather he/she can just look at the network by zooming in or zooming out. In this recipe, you are going to learn about another JavaScript plugin named Sigma.js, which is also available at Gephi Marketplace, that overcomes this limitation.

Getting ready

In order to use this plugin, we first need to install it in Gephi. To do so, follow these steps:

1. Run Gephi.
2. Navigate to **Tools** | **Plugins**.
3. Click on the **Available Plugins** tab and select **SigmaExporter** from the list of available plugins.
4. Hit **Install** and follow the wizard to install the SigmaExporter plugin in Gephi.
5. Select **Restart Now** once done.
6. Hit **Finish**. This installs the SigmaExporter plugin in Gephi.

How to do it...

The Sigma.js plugin for Gephi offers a really awesome way to generate beautiful, interactive visualizations. To generate one such visualization using the Les Misérables graph that is available in Gephi defaults, follow these steps:

1. Run Gephi and load the Les Misérables network.
2. Click on **File** in the menu bar.
3. In the drop-down list, select **Export**, followed by **Sigma.js template**.
4. In the **Sigma.js Export** window that opens, browse to the location in which you want the visualization to be exported to.
5. Enter the necessary information in the **Node**, **Edge**, and **Color** textboxes under **Legend**. These respectively represent the entity the nodes refer to, the relationships the edges depict, and the information that the color in the graph gives to the user.

This is how the dialog window looks:

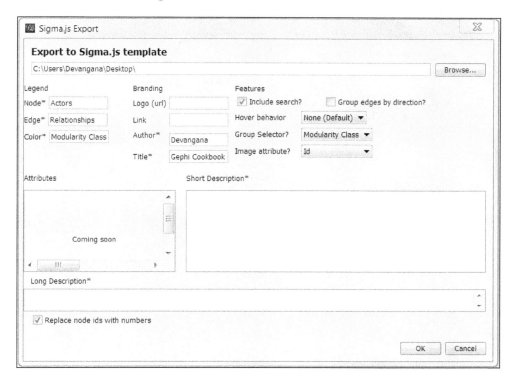

6.   Under **Branding**, enter the **Logo, Link, Author,** and **Title** information.

7.   Check the checkbox right next to **Include search** if you want the search option to be enabled in the final visualization.

8.   Set the remaining attributes and options accordingly.

9.   Enter the description for the network in the **Short Description** and/or **Long Description** textboxes.

10. Hit **OK** once done.

11. Navigate to the path where you exported the visualization. The files will typically be in a folder named network at the location.

12. Go inside that folder and open the `index.html` file in a Web browser, preferably Firefox. This is how the visualization will look in the browser.

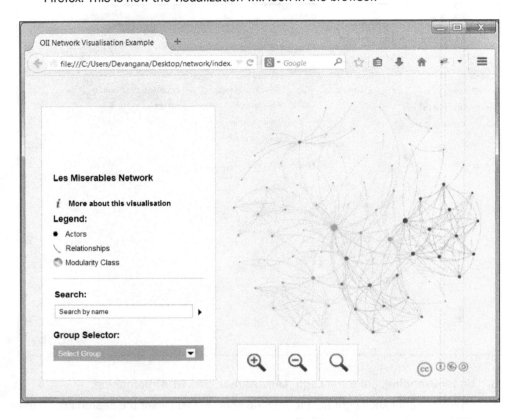

13. The **Legend** gives the information about the network that you entered during the export. If you want to know what a particular node represents, simply hover the mouse over it. It'll get highlighted and its name will be displayed adjacent to it.

14. If you want to select a group of nodes according to the group selector attribute that you selected during the export, click on the **Group Selector** drop-down menu and select the group of nodes that you want to highlight.

15. You can perform zoom in and zoom out on the network by using the buttons with magnifier icons that are displayed at the bottom of the screen.

16. You can get detailed information about a node and its neighbors by simply clicking on a node, which opens up a side panel that contains the details.

The following screenshot shows one such node being highlighted, with its details shown in the side panel:

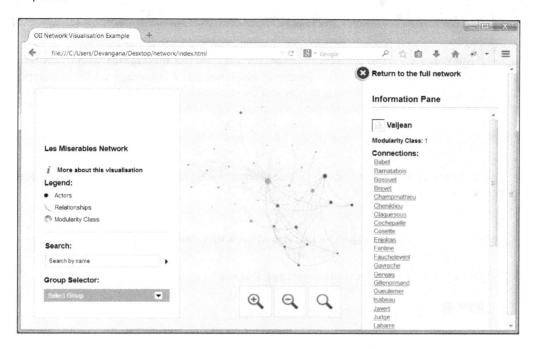

How it works...

Sigma.js is a JavaScript library that makes it easy to publish networks on Web pages. The Sigma.js plugin for Gephi allows its users to carry out the same task in Gephi. Using the Sigma.js plugin, you can export your network to a predefined Sigma.js template. You can have multiple ways of interacting with graphs: search, explanatory text, legends, group selection, and so on.

Sigma.js is a very useful and interactive plugin, but it consumes a lot of memory and hence it is not very useful with very large networks. For this reason, Sigma.js should be used with small networks or in cases where very large networks are involved, a powerful machine should be used.

See also

- ► `https://marketplace.gephi.org/plugin/sigmajs-exporter/` to learn more about Gephi's Sigma.js plugin
- ► A couple of interesting examples of visualizations using the Sigma.js plugin of Gephi at `http://blogs.oii.ox.ac.uk/vis/`

Describing complex network structures using GEXF

Since the beginning of this chapter, we have used the Les Misérables network for most of the recipes. If you noticed, on the **Welcome** screen, you click on **Les Miserables.gexf** to load the Les Misérables dataset in Gephi. Ever wondered what that **GEXF** extension is? In this recipe, you are going to learn what GEXF is and how you can leverage this format to construct really complex networks in a comparatively easy way. So let's get started.

How to do it...

The following steps give you an understanding of what the GEXF format is and how it is used to describe graphical structures:

1. Run Gephi.

2. From the **Welcome** screen, select **Les Miserables.gexf** to load the Les Misérables network into Gephi.

3. From the **Statistics** panel, run **Average Degree** to generate the degree value for all the vertices.

4. Go to the **File** option in the menu bar.

5. From the drop-down menu, select **Export**, followed by **Graph File**. This opens up an **Export** window, as shown in the following screenshot:

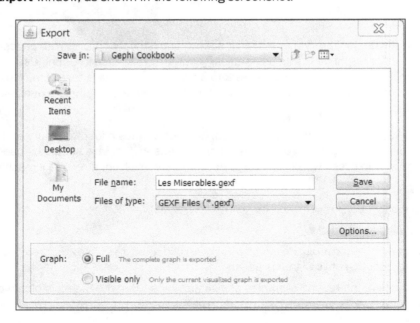

6. From the **Files of type** drop-down menu, select **GEXF Files (*.gexf)**.

7. Enter the filename for the file to be exported.

8. Click on **Options** and select the attributes from the graph that you want to be included in the exported file.

9. Click on **OK** once you have selected the options. Finally, click on **Save** to save the network data in the GEXF format.

10. Go to the location where you have saved this file and open it in an editor such as Notepad++. An editor such as Notepad++ instead of Notepad gives a better laydown to the text and hence it becomes easy to understand the structure of a file, such as the one we are looking at here. The following screenshot shows this exported data:

```xml
<?xml version="1.0" encoding="UTF-8"?>
<gexf xmlns="http://www.gexf.net/1.2draft" version="1.2" xmlns:viz="
http://www.gexf.net/1.2draft/viz" xmlns:xsi="
http://www.w3.org/2001/XMLSchema-instance" xsi:schemaLocation="
http://www.gexf.net/1.2draft http://www.gexf.net/1.2draft/gexf.xsd">
  <meta lastmodifieddate="2015-04-05">
    <creator>Gephi 0.8.1</creator>
    <description></description>
  </meta>
  <graph defaultedgetype="undirected" mode="static">
    <attributes class="node" mode="static">
      <attribute id="modularity_class" title="Modularity Class"
      type="integer"></attribute>
      <attribute id="degree" title="Degree" type="integer">
        <default>0</default>
      </attribute>
    </attributes>
    <nodes>
      <node id="0" label="Myriel">
        <attvalues>
          <attvalue for="modularity_class" value="0"></attvalue>
          <attvalue for="degree" value="10"></attvalue>
        </attvalues>
        <viz:size value="28.685715"></viz:size>
        <viz:position x="-266.82776" y="-299.6904"
        z="0.0"></viz:position>
        <viz:color r="91" g="91" b="245"></viz:color>
      </node>
```

How it works...

GEXF stands for **Graph exchange XML format**. GEXF is a language for describing complex networks in a simple XML-like schema. GEXF was developed by a group of people working on the Gephi project in 2007 and since then, it has transformed into a pretty mature form that can be used for real-world applications involving networks.

The GEXF file for a graph has information about the nodes and edges in the graph, their attributes, their properties, and the information about their dynamics. This information is present in a hierarchical format for easy and straightforward interpretation of the network, just by looking at the GEXF file. The structure of this file very much resembles the XML format. There are entities at one level with certain attributes and properties, and then there are entities that are encompassed within these entities, and so on. This clearly defines the relationship structure that the nodes in a graph may have within themselves in the graph.

In the previous screenshot of the data that we had in the previous section, you will notice some metadata information about the exported data in the beginning, such as Gephi's version, last modified date, and so on. The data of interest for us here, though, is the nodes and their attribute information. For example, the node at the topmost hierarchy here is a graph node that contains information about graph properties, such as whether the graph is directed or undirected. It then contains a node under it called the attribute node that has information about generic properties of the nodes in the network data. In the same hierarchy, it contains many children of nodes that represent each specific node in the network and their properties, such as the values for each of the attributes defined under the attribute node, position in the graph, color and so on. You can try running some more metrics and filters on the graph and then explore the hierarchical structure of the GEXF data for that graph.

See also

▸ `http://gexf.net/format/` to learn more about the GEXF language

▸ A pretty good example involving file export into GEXF from Gephi at `https://wiki.issuecrawler.net/Issuecrawler/GexfExport`

Generating world maps

So far in this chapter, we have learned about plugins that help in representing and exporting the graphs in different formats and allow users to build interactive applications with them. In this recipe, you are going to learn about a very interesting plugin called **Map of Countries**, which is available at Gephi Marketplace. This plugin comes in handy when you want to design or draw maps of countries in Gephi and do some more exploration on top of them.

Getting ready

In order to start playing around with this plugin, you first need to install it. Run Gephi, go to **Tools | Plugins**, and from there install the Map of Countries plugin from the list of available plugins, in a similar way to how you installed the plugins already covered in this chapter.

How to do it...

The following steps discuss various features that the Map of Countries plugin has to offer and what you can achieve with those features:

1. Run Gephi. In the **Welcome** window, select **New Project**.
2. In the **Layout** panel, select **Map of Countries** from the list of layouts.
3. To draw the entire world map, select **World** from the **Country** drop-down menu in the options under the **Layout** panel.
4. Hit **Run** to generate a world map, as shown in the following screenshot:

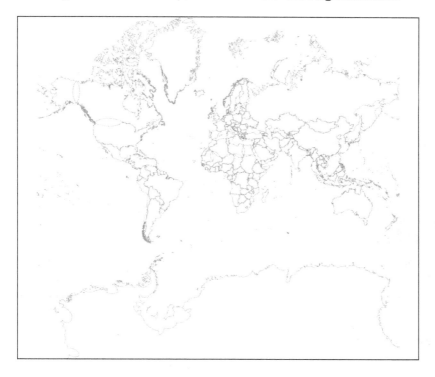

5. To generate a different type of projection, click on the drop-down menu for **Projection** and select the required option.
6. You can optionally choose to draw the map of only a specific country. To do so, select the country from the drop-down menu for **Country**.

7. Hitting **Run** will generate a graph similar to the one shown in the following screenshot, depending on the country chosen:

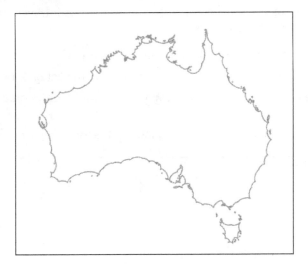

8. You can also draw the map of a subregion of the world by selecting the appropriate option, such as **Eastern Africa**, from the **Subregion** drop-down menu.

9. You can draw the map of a region by selecting an appropriate option, such as **Asia**, from the **Region** drop-down menu.

How it works...

We introduced the concept of map projections in this recipe. A map projection is a representation of the longitudes and latitudes of places on the surface of the earth, which is three-dimensional, onto a two-dimensional surface. The projections obviously tend to distort the actual surface and hence, keeping in mind different factors, different varieties of projections exist. Some of these projections are implemented in the Map of Countries plugin and are available ready to use.

The plugin uses the World Borders Dataset, which is available at `http://thematicmapping.org/downloads/world_borders.php`. The dataset contains the regions and countries that are defined in the form of polygons, and each polygon has its corresponding longitude(s) and latitude(s) defined. This information is used by Gephi to draw the map of a country or a region.

See also

▶   `https://marketplace.gephi.org/plugin/maps-of-countries/` for details about the Map of Countries plugin

> ▶ `http://en.wikipedia.org/wiki/Map_projection` to know more about map projections

> ▶ `http://en.wikipedia.org/wiki/List_of_map_projections` for a good and crisp documentation of the map projections

> ▶ `http://www.viewsoftheworld.net/?p=752` to know more about map projections

Performing social network analysis

The field of social network analysis is one of the most sought-after research fields in the multidisciplinary worlds of sociology and computer science today. A huge number of researchers across the world are working in this field. With the advent of social networks, the way we connect with other people in the world has totally changed from the way we used to connect before. The world has become much smaller and the ties have probably grown stronger. A lot of interesting behavioral and organizational aspects come into play when one is researching social networks. A lot of the core features of Gephi will aid in the study of these social networks. Gephi Marketplace has a plugin that lets the users do even more. The plugin is called Social Network Analysis and was built by Jaroslav Kuchar from Czech Technical University in Prague. In this recipe, we are going to cover this plugin and learn what features it offers.

Getting ready

In order to start with this plugin, first install it from Gephi Marketplace in a similar manner as you installed the other plugins covered in this chapter. The plugin is known as SNAMetricsPlugin.

How to do it...

Once you have installed this plugin, follow these steps to explore various features of it:

1. Run Gephi and load the Les Misérables network.

2. From the **Statistics** panel, click on **Run** next to **Erdös Number** in the **Statistics** panel. This opens up a window with the list of nodes in the Les Misérables network.

3. Select the node for which you want to compute the Erdös number and hit **OK**. This computes the Erdös number for that particular node in the Les Misérables network.

4. Under the **Partition** panel, click on the **Nodes** tab.

5. Click on the button with the refresh symbol on it to populate the list of partition parameters.

6. Select **Erdös Number** from the drop-down list. Hit **Apply** to generate a coloring scheme for the Les Misérables network based on the Erdös number of the nodes, with respect to the selected node, similar to the network shown in the following screenshot:

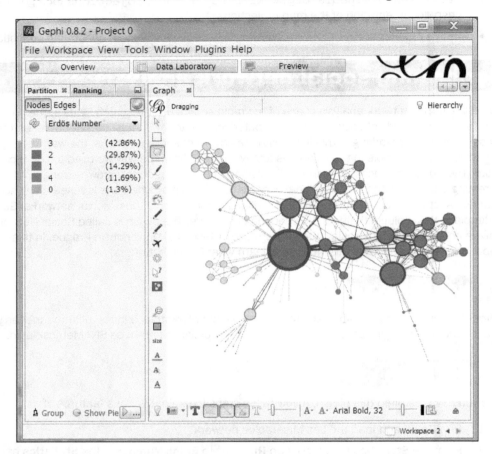

7. The plugin also allows you to find out the **Neighborhood Overlap** and **Embeddedness**. To do that, click on the **Run** button next to the **Neighborhood Overlap, Embeddedness** option under the **Statistics** panel.

8. Hit **OK** in the window that opens up. This opens up the **Overlap Neighborhood and Embeddedness Report** window, as shown in the following screenshot:

How it works...

Social network analysis, as mentioned before, is one of the hottest research topics in the field of computer science right now. There are some very interesting problems that have been solved in the past and are being solved right now by thousands of researchers from all around the world in this field. The field is vast, encompassing various domains ranging from computer science, economics, behavioral sciences, political sciences, and many more. This plugin tries to give insight into a very particular metric used quiet often in the field of social network analysis. This metric is known as the Erdös number. The tool also helps to determine yet another metric called **Neighborhood Overlap and Embeddedness**. Let's discuss these three metrics in detail.

The Erdös number is named after a very famous and influential mathematician named Paul Erdös. Paul Erdös published a huge number of papers in his lifetime and hence this number was a tribute to him. The Erdös number of a person is the number of hops between Paul Erdös and that person, measured by the authorship of published research papers. This means that, if a person has published one or more research papers with Paul Erdös, then the Erdös number of that person is 1. If a person has published research paper(s) with someone who has directly published paper(s) with Paul Erdös, then that person's Erdös number is 2, and so on.

The embeddedness of an edge in a network is the number of nodes that are neighbors of both the nodes that are connected via that edge. Neighborhood overlap of an edge is the embeddedness of the edge divided by the total number of neighbors of the nodes of that edge. The average neighborhood overlap of a graph is the average of the neighborhood overlap values for all the edges in the graph. Similarly, the average embeddedness of a graph is the average of the embeddedness values for all the edges in the graph.

There's more...

Social network analysis is a huge field and has a large number of application areas. The Internet is full of resources on this topic. There are some really good, highly-recommended resources that you can go through if you are interested in learning about this area.

See also

▸ The Social Network Analysis course by Lada Adamic of University of Michigan on Coursera at `https://www.coursera.org/course/sna`

▸ *Social Network Analysis: Methods and Applications* by Stanley Wasserman and Katherine Faust

▸ *Social Network Analysis* by John Scott

▸ *Networks, Crowds and Markets: Reasoning About a Highly Connected World* by David Easley and Jon Kleinberg

▸ `http://wwwp.oakland.edu/enp/` to learn about the Erdös number project in detail

Index

Thank you for buying
Gephi Cookbook

About Packt Publishing

Packt, pronounced 'packed', published its first book, *Mastering phpMyAdmin for Effective MySQL Management*, in April 2004, and subsequently continued to specialize in publishing highly focused books on specific technologies and solutions.

Our books and publications share the experiences of your fellow IT professionals in adapting and customizing today's systems, applications, and frameworks. Our solution-based books give you the knowledge and power to customize the software and technologies you're using to get the job done. Packt books are more specific and less general than the IT books you have seen in the past. Our unique business model allows us to bring you more focused information, giving you more of what you need to know, and less of what you don't.

Packt is a modern yet unique publishing company that focuses on producing quality, cutting-edge books for communities of developers, administrators, and newbies alike. For more information, please visit our website at www.packtpub.com.

About Packt Open Source

In 2010, Packt launched two new brands, Packt Open Source and Packt Enterprise, in order to continue its focus on specialization. This book is part of the Packt open source brand, home to books published on software built around open source licenses, and offering information to anybody from advanced developers to budding web designers. The Open Source brand also runs Packt's open source Royalty Scheme, by which Packt gives a royalty to each open source project about whose software a book is sold.

Writing for Packt

We welcome all inquiries from people who are interested in authoring. Book proposals should be sent to author@packtpub.com. If your book idea is still at an early stage and you would like to discuss it first before writing a formal book proposal, then please contact us; one of our commissioning editors will get in touch with you.

We're not just looking for published authors; if you have strong technical skills but no writing experience, our experienced editors can help you develop a writing career, or simply get some additional reward for your expertise.

Mastering Gephi Network Visualization

ISBN: 978-1-78398-734-4   Paperback: 378 pages

Produce advanced network graphs in Gephi and gain valuable insights into your network datasets

1. Build sophisticated interactive network graphs using advanced Gephi layout features.

2. Master Gephi statistical and filtering techniques to easily navigate through even the densest network graphs.

3. An easy-to-follow guide introducing you to Gephi's advanced features, with step-by-step instructions and lots of examples.

Network Graph Analysis and Visualization with Gephi

ISBN: 978-1-78328-013-1   Paperback: 116 pages

Visualize and analyze your data swiftly using dynamic network graphs built with Gephi

1. Use your own data to create network graphs displaying complex relationships between several types of data elements.

2. Learn about nodes and edges, and customize your graphs using size, color, and weight attributes.

3. Filter your graphs to focus on the key information you need to see and publish your network graphs to the Web.

Please check **www.PacktPub.com** for information on our titles

Learning Neo4j

ISBN: 978-1-84951-716-4   Paperback: 222 pages

Run blazingly fast queries on complex graph datasets with the power of the Neo4j graph database

1. Get acquainted with graph database systems and apply them in real-world use cases.

2. Get started with Neo4j, a unique NoSQL database system that focuses on tackling data complexity.

3. A practical guide filled with sample queries, installation procedures, and useful pointers to other information sources.

Google Visualization API Essentials

ISBN: 978-1-84969-436-0   Paperback: 252 pages

Make sense of your data: make it visual with the Google Visualization API

1. Wrangle all sorts of data into a visual format, without being an expert programmer.

2. Visualize new or existing spreadsheet data through charts, graphs, and maps.

3. Full of diagrams, core concept explanations, best practice tips, and links to working book examples.

Please check **www.PacktPub.com** for information on our titles